Workbook

Modern Refrigeration and Air Conditioning

by

A. F. Bracciano
D. C. Bracciano
G. M. Bracciano

Publisher
The Goodheart-Willcox Company, Inc.
Tinley Park, IL
www.g-w.com

The Goodheart-Willcox Company, Inc. Brand Disclaimer: Brand names, company names, and illustrations for products and services included in this text are provided for educational purposes only and do not represent or imply endorsement or recommendation by the author or the publisher.

The Goodheart-Willcox Company, Inc. Safety Notice: The reader is expressly advised to carefully read, understand, and apply all safety precautions and warnings described in this book or that might also be indicated in undertaking the activities and exercises described herein to minimize risk of personal injury or injury to others. Common sense and good judgment should also be exercised and applied to help avoid all potential hazards. The reader should always refer to the appropriate manufacturer's technical information, directions, and recommendations; then proceed with care to follow specific equipment operating instructions. The reader should understand these notices and cautions are not exhaustive.

The publisher makes no warranty or representation whatsoever, either expressed or implied, including but not limited to equipment, procedures, and applications described or referred to herein, their quality, performance, merchantability, or fitness for a particular purpose. The publisher assumes no responsibility for any changes, errors, or omissions in this book. The publisher specifically disclaims any liability whatsoever, including any direct, indirect, incidental, consequential, special, or exemplary damages resulting, in whole or in part, from the reader's use or reliance upon the information, instructions, procedures, warnings, cautions, applications, or other matter contained in this book. The publisher assumes no responsibility for the activities of the reader.

The Goodheart-Willcox Company, Inc. Internet Disclaimer: The Internet resources and listings in this Goodheart-Willcox Publisher product are provided solely as a convenience to you. These resources and listings were reviewed at the time of publication to provide you with accurate, safe, and appropriate information. Goodheart-Willcox Publisher has no control over the referenced websites and, due to the dynamic nature of the Internet, is not responsible or liable for the content, products, or performance of links to other websites or resources. Goodheart-Willcox Publisher makes no representation, either expressed or implied, regarding the content of these websites, and such references do not constitute an endorsement or recommendation of the information or content presented. It is your responsibility to take all protective measures to guard against inappropriate content, viruses, or other destructive elements.

Cover images: Arkema, Inc.; Stride Tool Inc.; Danfoss; Tempstar
Back cover images: Ritchie Engineering Co., Inc. – YELLOW JACKET Products Division;
Emerson Climate Technologies

Introduction

This workbook is designed for use with the text, *Modern Refrigeration and Air Conditioning*. Each workbook chapter should be completed after reading the corresponding text chapter. The workbook serves as an open book quiz on the contents of the textbook.

Each chapter of the workbook includes a variety of question types. The types of questions include multiple choice, true or false, matching, and short answer. Some questions involve calculations. For these questions, be sure to show your work in the space provided.

The workbook questions are grouped by textbook section for easy usability. Each workbook chapter ends with a *Critical Thinking* section. The questions in these sections will allow you to consider and apply the knowledge you have gained from the chapter content.

Reading *Modern Refrigeration and Air Conditioning* and using this workbook will help you acquire a working knowledge of the principles of refrigeration and air conditioning and their application. Answering the questions for each chapter will help you master the technical knowledge presented in the text.

Table of Contents

CHAPTER 1

Careers and Certification

Name _____

Date _____ Class _____

Carefully study the chapter and then answer the following questions.

1.2 Career Planning

_____ 1. A career is _____.
- A. a job that requires a college degree
- B. a job that you do for more three years
- C. any job that provides health insurance
- D. a series of employment opportunities, each requiring increasing skill and knowledge

_____ 2. *True or False?* The goal in a career is to move into positions requiring greater knowledge and skill.

_____ 3. Two types of service work are maintenance and _____.
- A. system repair
- B. installation of new equipment in existing structures
- C. installation in new structures
- D. None of the above.

_____ 4. *True or False?* Residential HVACR installation work occurs in only new construction.

_____ 5. *True or False?* Each of the three career pathways in the Architecture and Construction includes HVACR-related occupations.

_____ 6. Changing filters in and monitoring the cooling effectiveness of a refrigeration system in a grocery store would be completed by someone in an occupation found in the _____ pathway.
- A. construction
- B. design/preconstruction
- C. maintenance/operations
- D. None of the above.

_____ 7. Which of the following occupations is likely to spend the most time using computer software to determine air-conditioning loads for a commercial building?
- A. HVACR drafter
- B. HVACR engineer
- C. HVACR installer
- D. HVACR service technician

_____ 8. *True or False?* An estimator calculates the cost of the equipment required for a project but cannot estimate the labor cost until after the project is completed.

_____ 9. *True or False?* Companies avoid energy audits because the changes suggested by the energy auditor normally result in increased electricity and gas costs in the future.

_____ 10. *True or False?* For new residential construction, the building inspection begins after the HVAC systems are installed.

_____ 11. An HVACR instructor may be employed by _____.
   A. an HVACR equipment manufacturer
   B. a technical school or community college
   C. a training organization
   D. All of the above.

_____ 12. Which of the following occupations is likely to be most familiar with local building codes?
   A. Building inspector
   B. Energy auditor
   C. Estimator
   D. HVACR drafter

_____ 13. Over the next five years, the number of available positions for HVACR technicians is expected to _____.
   A. decrease significantly (more than 20%)
   B. decrease slightly (less than 5%)
   C. increase significantly (more than 20%)
   D. increase slightly (less than 5%)

1.3 Beginning Your Career Search

_____ 14. *True or False?* The contacts and network you develop as a student and apprentice may lead to career opportunities.

_____ 15. A career website may allow you to _____.
   A. post your résumé
   B. receive updates of new job postings
   C. view job postings from companies
   D. All of the above.

_____ 16. Which of the following provides a summary of your educational background and work accomplishments?
   A. Cover letter
   B. Career passport
   C. Résumé
   D. All of the above.

_____ 17. *True or False?* In your cover letter, you should request further discussion or an interview.

_____ 18. When completing a job application, _____.
   A. be truthful
   B. complete all questions
   C. write neatly
   D. All of the above.

Name _____

_____ 19. When dressing for an interview, select an outfit that is _____ the clothing typically worn by someone in the position.

    A. identical to
    B. less formal than
    C. more formal than
    D. None of the above.

1.4 Success in the Workplace

_____ 20. *True or False?* Your appearance will affect your level of success in the workplace.

_____ 21. The habit of identifying what must be done and doing it without being told is called _____.

    A. leadership
    B. punctuality
    C. taking initiative
    D. teamwork

_____ 22. Arriving on time for work is an example of _____.

    A. leadership
    B. punctuality
    C. taking initiative
    D. teamwork

_____ 23. *True or False?* Exhibiting strong leadership skills will help you advance in your career.

_____ 24. *True or False?* Complaining about work with your coworker is a good way to show leadership.

_____ 25. *True or False?* Conflicts and disagreements never occur in a professional workplace.

_____ 26. Having good _____ skills is critical to being an effective member of a team.

    A. communication
    B. grooming
    C. punctuality
    D. time management

1.6 Certification

_____ 27. A certification shows that you _____.

    A. are a good worker
    B. are ready for a full-time position
    C. have attained a level of competence in a specific set of topics
    D. have received a passing grade for your course

_____ 28. All HVACR technicians who work with refrigerants that have the potential to harm the environment must be certified by the EPA in accordance with Section 608 of the _____.

    A. Clean Air Act
    B. Environmental Protection Act
    C. Kyoto Protocol
    D. National Industrial Recovery Act

Critical Thinking

29. Why is lifelong learning critical in the HVACR industry?

_____

_____

_____

_____

_____

30. Find job postings for HVACR positions on the Internet. List five positions and the companies that are hiring.

_____

_____

_____

_____

_____

31. Select one general career website and one HVACR-specific website with job postings. Compare and contrast the HVACR positions listed on the sites. Which will be a better resource for finding HVACR positions?

_____

_____

_____

_____

_____

_____

CHAPTER 2
Safety

Name _____

Date _____ **Class** _____

Carefully study the chapter and then answer the following questions.

2.1 Safety and the Government

_____ 1. The national law that covers workplace safety is ____.

    A. ADA

    B. EPA

    C. OSHA

    D. SDS

_____ 2. Companies are required to report ____.

    A. accident involving three or more hospitalized injuries

    B. every fatality

    C. permanently disabling injury

    D. All of the above.

2.2 Hazard Assessment

_____ 3. A hazard is anything that ____.

    A. creates a potential for harm

    B. has resulted in an injury

    C. is too dangerous for the jobsite

    D. requires safety glasses and hearing protection

_____ 4. *True or False?* It is best to assume a power source is live until you have locked it out.

_____ 5. Lockout/tagout procedures ____.

    A. ensure that electrical circuits remain de-energized while they are being worked on

    B. identify all workers currently on the work site

    C. prevent tool theft

    D. prevent unauthorized workers from entering a hazardous area

_____ 6. *True or False?* Under no circumstances can electrical equipment be energized while it is worked on.

Match the following classes of fires with the phrase that best describes them.

_____ 7. Flammable liquids, such as gasoline and oil

_____ 8. Combustible metals, such as magnesium and lithium

_____ 9. Ordinary combustibles, such as wood, paper, and textiles

_____ 10. Electrical fires, such as motors and switches

A. Class A

B. Class B

C. Class C

D. Class D

Match the hazard symbols with the corresponding hazard type.

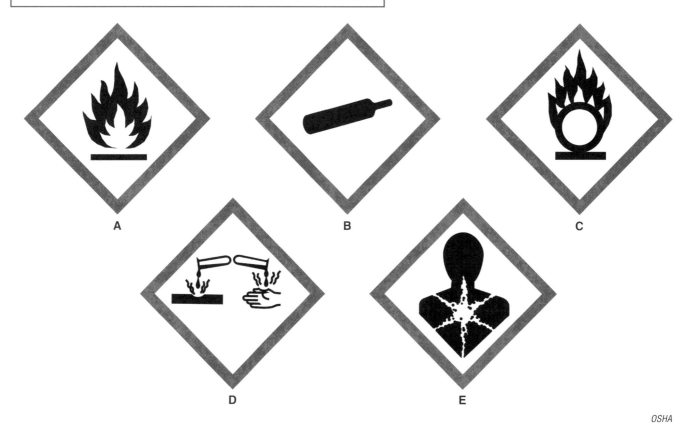

A   B   C

D   E

OSHA

_____ 11. Substance can cause chemical burns

_____ 12. Flammable substance

_____ 13. Substance is an oxidizer (promotes combustion)

_____ 14. Harmful to humans

_____ 15. Cylinder with gas under pressure

Name _____

_____ 16. *True or False?* The safest way to move a refrigerant cylinder is to lay it on its side and roll it.

_____ 17. Refillable refrigerant cylinders should not be filled more than _____.

    A. 50%
    B. 80%
    C. 100%
    D. 150%

_____ 18. *True or False?* When a pressure relief valve in a cylinder opens, all of the refrigerant is released from the cylinder.

_____ 19. Based on ASHRAE Standard 34, which refrigerant classification would indicate greatest flammability?

    A. Class A1
    B. Class A2
    C. Class A3
    D. All are equally flammable.

_____ 20. A stationary refrigerant detector indicates when _____.

    A. pressure in a refrigerant cylinder is too high
    B. refrigerant level in a system is too low
    C. refrigerant level in a system is too high
    D. refrigerant vapor level in the air is too high

_____ 21. *True or False?* The Globally Harmonized System (GHS) is a standard system for labeling chemicals.

Identify each hazard with its hazard type. Each answer will be used more than once.

_____ 22. Live electrical wires

_____ 23. Waste building materials scattered on the ground

_____ 24. High concentration of refrigerant gas in the air

_____ 25. Exposure to extremely hot and humid weather

_____ 26. Dust

_____ 27. Asbestos fibers

_____ 28. High pressure refrigerant leak

_____ 29. Improper tool operation

    A. Health hazard

    B. Physical hazard

2.3 Personal Protective Equipment (PPE)

_____ 30. *True or False?* A hard hat should always be worn on a construction site.

Match the photos of personal protective equipment with the correct name.

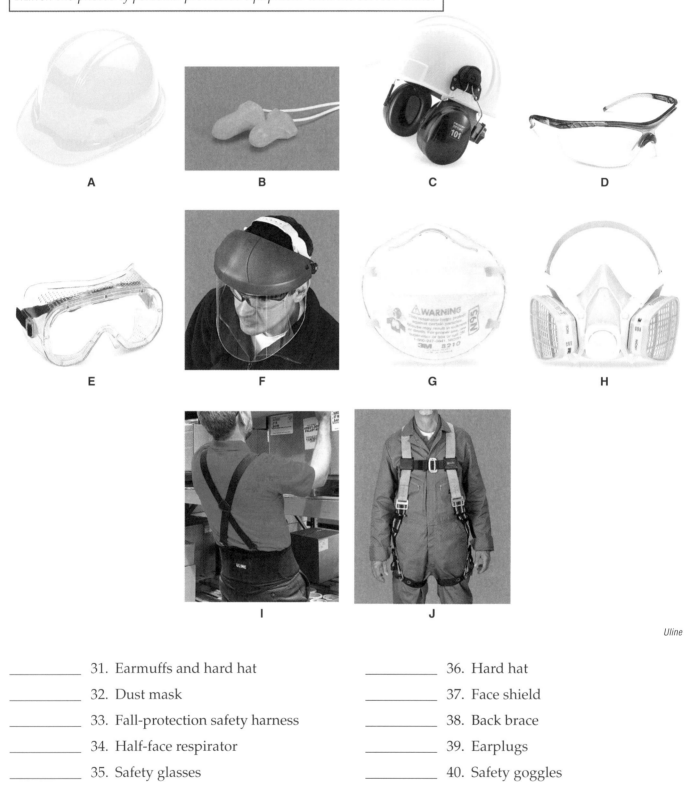

A

B

C

D

E

F

G

H

I

J

Uline

_____ 31. Earmuffs and hard hat

_____ 32. Dust mask

_____ 33. Fall-protection safety harness

_____ 34. Half-face respirator

_____ 35. Safety glasses

_____ 36. Hard hat

_____ 37. Face shield

_____ 38. Back brace

_____ 39. Earplugs

_____ 40. Safety goggles

Name _____

2.4 Safe Work Practices

_____ 41. When lifting heavy objects, _____.
   A. lift with your legs, not your back
   B. use a hand truck or dolly
   C. wear a back brace for additional support
   D. All of the above.

_____ 42. When an extension ladder is placed against a roof, the top of the ladder should extend a minimum of _____ beyond the support point.
   A. 6″
   B. 1′
   C. 3′
   D. 5′

_____ 43. Fall protection is required when there is a hazard of a drop greater than _____.
   A. 2′
   B. 6′
   C. 12′
   D. 20′

Critical Thinking

44. Who is responsible for your safety on the job? Explain your answer.

_____

_____

_____

_____

_____

45. List at least one example of each of the seven hazards you might encounter while performing a service call on a commercial air-conditioning system.

   Electrical: _____

   Fire: _____

   Temperature: _____

   Pressure: _____

   Refrigerants: _____

   Chemical: _____

   Breathing: _____

46. List three potential hazards when working with a portable extension ladder.

_____

_____

_____

_____

CHAPTER 3

Service Calls

Name _____

Date _____ Class _____

> *Carefully study the chapter and then answer the following questions.*

3.1 Servicing

_____ 1. Service required to fix a system that is not operating correctly describes _____.
   A. customer service
   B. maintenance
   C. repair
   D. upgrades

_____ 2. Service performed regularly to reduce the likelihood of a future breakdown and to minimize any reduction of system performance describes _____.
   A. customer service
   B. maintenance
   C. repair
   D. upgrades

3.2 Troubleshooting

> *Match each troubleshooting step description with the correct step number in proper order.*

_____ 3. Identify a specific remedy for the problem.

_____ 4. Determine the possible cause from the problem's description.

_____ 5. Obtain a description of the problem and a list of recent repairs from the owner.

_____ 6. Verify the suspected cause using pinpoint tests.

A. Step 1

B. Step 2

C. Step 3

D. Step 4

_____   7. Many HVAC units use LEDs to flash a _____, which indicates that there is an equipment malfunction.
   A. callback
   B. remedy
   C. standard procedure
   D. trouble code

_____   8. A hasty diagnosis and temporary fix to a system are likely to result in a _____.
   A. big tip
   B. callback
   C. job promotion
   D. trouble code

_____   9. *True or False?* A temporary fix always uncovers the root cause of a problem.

_____   10. If a system is low on refrigerant, a technician should check for _____.
   A. bad electrical wiring practices
   B. overcooling problems
   C. refrigerant leaks
   D. restricted airflow

3.3 Customer Service

_____   11. During a service call, information provided by the customer concerning previous problems with the equipment should be written on _____.
   A. the back of a take-out menu
   B. a dirty napkin
   C. the palm of your hand
   D. a work order or service contract

_____   12. *True or False?* A technician's appearance and conduct contribute to the company image.

_____   13. *True or False?* An estimate does *not* need to factor in company overhead costs, since they are unrelated to the customer's service call.

_____   14. *True or False?* Two features of a service contract that often appeal to purchasers include a 24-hour availability clause and an absolute guarantee of work done.

_____   15. The way a business interacts with customers is called _____.
   A. customer relations
   B. damage control
   C. HAZMAT protocol
   D. troubleshooting

Name _____

Critical Thinking

> *Use the troubleshooting charts located in the textbook and the Appendix to answer Questions 16–20.*

16. A customer with a light commercial refrigerating unit reports that the unit operates long or continuously. List three possible causes.

_____

_____

_____

17. The owner of a hermetic refrigeration system complains that the temperature inside the refrigerator is too high. List two possible remedies to the problem.

_____

_____

18. The owner of a forced-air heating system complains that the unit produces unsatisfactory heat because it runs for only a short period and then stops. List the four possible causes.

_____

_____

_____

_____

19. The owner of an industrial food storage system reports that the motor hums but will not start. What are three possible causes?

_____

_____

_____

20. List three things that a technician typically checks when performing scheduled maintenance on a residential forced-air heating and cooling unit.

_____

_____

_____

_____

Notes

CHAPTER 4

Energy and Matter

Name _____

Date _____ **Class** _____

Carefully study the chapter and then answer the following questions.

4.1 Systems of Measurement

_____ 1. The base unit for mass in the US Customary measurement system is _____.

A. foot (ft)
B. gram (g)
C. meter (m)
D. pound (lb)

_____ 2. The base unit for distance in the US Customary measurement system is _____.

A. foot (ft)
B. gram (g)
C. meter (m)
D. pound (lb)

_____ 3. The base unit for mass in the SI measurement system is _____.

A. foot (ft)
B. gram (g)
C. meter (m)
D. pound (lb)

_____ 4. The base unit for distance in the SI measurement system is _____.

A. foot (ft)
B. gram (g)
C. meter (m)
D. pound (lb)

4.2 Matter and Energy

_____ 5. The following are the three physical states of a substance, *except* _____.

A. force
B. gas
C. liquid
D. solid

_____   6. *True or False?* Energy can change from one form to another.

_____   7. Energy can be classified into which two categories?
   A. Electrical and emotional.
   B. Integral and proportional.
   C. Kinetic and potential.
   D. Material and thermal.

4.3 Mass and Weight

_____   8. As the amount of a substance increases, the force of gravity acting on it _____.
   A. decreases proportionally
   B. increases proportionally
   C. is not affected at all
   D. All of the above.

_____   9. The term *mass* is used to express that the _____ of material is the same, regardless of the change in the force of gravity.
   A. heat content
   B. magnetism
   C. quality
   D. quantity

4.4 Density

_____   10. Density is best described as meaning the _____.
   A. dimensions of a substance as measured linearly
   B. force a substance exerts on gravity
   C. gravitational force exerted on a substance
   D. mass per unit of volume of a substance

_____   11. The ratio of the mass of a certain volume of a gas as compared to the mass of an equal volume of hydrogen is called _____.
   A. relative density
   B. specific gravity
   C. specific volume
   D. None of the above.

_____   12. The volume of a specific amount of gas under standard conditions is referred to as _____.
   A. relative density
   B. specific gravity
   C. specific volume
   D. None of the above.

Name _____

4.5 Force, Work, and Power

_____ 13. Energy applied to matter that causes a change in the matter's velocity is called _____.

   A. adiabatic compression
   B. power
   C. force
   D. work

_____ 14. How much work (in foot-pounds) is required to move a block 10′ if the force is 200 lb?

   A. 20 ft-lb
   B. 190 ft-lb
   C. 210 ft-lb
   D. 2,000 ft-lb

_____ 15. Calculate the power in kilowatts (kW) required to lift a 500 kg mass 12 m in one second. Show your calculations.

   A. 41.66 kW
   B. 58.5 kW
   C. 408.3 kW
   D. 6000h kW

4.6 Heat

_____ 16. *True or False?* Heat always flows from a substance at a lower temperature to a substance at a higher temperature.

_____ 17. The absolute scale that uses the same divisions as the Fahrenheit scale is the _____ scale.

   A. Celsius
   B. Kelvin
   C. Rankine
   D. None of the above.

_____ 18. The temperature of the air surrounding an object is _____ temperature.

   A. ambient
   B. local
   C. neighboring
   D. regional

_____ 19. A Btu is the amount of heat required to raise one pound of water _____.

   A. by 1°C
   B. by 1°F
   C. to its boiling point
   D. to a level of 10°F superheat

Match heat transfer method with the appropriate illustration.

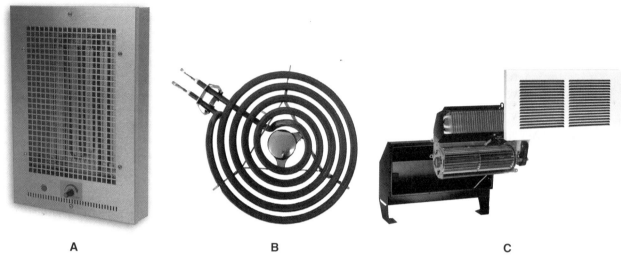

A   B   C

_____ 20. Conduction

_____ 21. Convection

_____ 22. Radiation

_____ 23. Poor conductors of heat are known as _____.
   A. conductors
   B. convectors
   C. insulators
   D. semiconductors

_____ 24. The highest temperature at which a substance may liquefy, regardless of pressure, is called its _____.
   A. critical temperature
   B. latent heat of fusion
   C. specific enthalpy
   D. specific heat capacity

_____ 25. The following statements describe sensible heat, *except* _____.
   A. heat that can be measured as superheat or subcooling
   B. heat that causes a change of physical state
   C. heat that causes temperature rise
   D. heat that causes temperature drop

Name _____

4.7 Measuring Refrigeration Effect

_____ 26. A ton of refrigeration is equal to the heat energy required to melt one ton of ice in _____.

    A. 1 minute

    B. 1 hour

    C. 12 hours

    D. 24 hours

_____ 27. How many Btu per hour is equivalent to one ton of refrigeration?

    A. 200 Btu/hr

    B. 12,000 Btu/hr

    C. 144,000 Btu/hr

    D. 288,000 Btu/hr

Critical Thinking

_____ 28. Calculate the amount of heat (in Btu) needed to raise the temperature of 2.5 lb of glass from 45°F to 350°F. The specific heat capacity of glass is 0.187 Btu/lb°F. Show your calculations.

_____ 29. Which of the following has more total enthalpy: 3 kg of 20% salt brine at 65°C or 5 kg of 20% salt brine at 50°C, assuming 0 enthalpy at 0°C? The specific heat capacity of 20% salt brine is 3.56 kJ/kg°C. Show your calculations.

_____ 30. An air conditioner for a single-family residence has a capacity of 1,008,000 Btu per day. Calculate the tonnage of this system. How many tons of refrigeration is it designed to handle? Show your calculations.

CHAPTER **5**

Gases

Name _____

Date _____ Class _____

2

> *Carefully study the chapter and then answer the following questions.*

5.1 Volume

_____ 1. As a substance is heated, its volume tends to _____.

    A. decrease

    B. increase

    C. get louder

    D. remain unchanged

_____ 2. In most cases, solids and liquids are _____.

    A. compressible

    B. noncompressible

    C. All of the above.

    D. None of the above.

5.2 Pressure

_____ 3. Often expressed in psi, pressure is the term for force per unit of _____.

    A. area

    B. length

    C. mass

    D. volume

_____ 4. Absolute pressure is gauge pressure plus _____ pressure.

    A. ambient

    B. atmospheric

    C. system

    D. vacuum

_____ 5. Pascal's law states that pressure applied on a confined fluid is _____.

    A. cooler the lower the pressure

    B. hotter the higher the pressure

    C. transmitted equally and undiminished

    D. repelled

_____ 6. Refrigerant passing through a compressor quickly undergoes _____ compression, which means it is compressed without losing heat to its surroundings.
   A. absolute
   B. adiabatic
   C. ambient
   D. perfect

_____ 7. _True or False?_ The temperature at which a liquid boils depends on its pressure.

| _Match the proper abbreviations for each unit listed._ |

_____ 8. Pound

_____ 9. Pounds per square inch gauge

_____ 10. Pounds per square inch absolute

_____ 11. Watt

_____ 12. Inches of mercury

_____ 13. British thermal unit

_____ 14. Degrees Fahrenheit

_____ 15. Degrees Celsius

_____ 16. Joule

A. Btu

B. °C

C. °F

D. in. Hg

E. J

F. lb

G. psia

H. psig

I. W

5.3 Gas Laws

_____ 17. According to Boyle's law, if the temperature of a gas is held constant but its volume is doubled, its pressure will _____.
   A. double
   B. quadruple
   C. reduce by half
   D. reduce by one-quarter

_____ 18. According to Charles' law, if the pressure of a gas is held constant but its temperature is increased, its volume will _____.
   A. decrease
   B. increase
   C. remain the same
   D. get louder

_____ 19. According to Gay-Lussac's law, if the volume of a gas is held constant but its temperature is increased, its pressure will _____.
   A. decrease
   B. increase
   C. remain the same
   D. burst the vessel

Name _____

_____ 20. According to Dalton's law, to determine the total pressure of a mixture of oxygen and hydrogen, you must _____ the pressures of each of the two gases.

A. add
B. multiply
C. square
D. subtract

_____ 21. Avogadro's law states that two different gases at the same temperature and pressure would occupy the same volume, even though each gas has a different _____.

A. adiabatic value
B. heat content
C. mass
D. physical state

5.4 Saturated Vapor

_____ 22. A saturated vapor is in balance such that some vapor will _____ if temperature decreases or pressure increases.

A. condense
B. evaporate
C. subcool
D. superheat

_____ 23. In a properly operating refrigeration system, saturated vapor would most likely be found in the _____.

A. compressor and condenser
B. evaporator and condenser
C. liquid line and compressor
D. suction line and liquid line

5.5 Basic Processes That Provide Cooling Effect

_____ 24. *True or False?* In air exchange, heat transfers from warm air to a cooler surface without causing a state change.

_____ 25. *True or False?* In pressure change cooling effect, the temperature of a fluid in a confined volume decreases as pressure increases.

_____ 26. *True or False?* When matter changes state from liquid to solid, the matter is absorbing heat.

Critical Thinking

27. Determine the pressure applied to a 5″ × 5″ area by 525 lb of force. Show your calculations.

28. Explain how the principles of Gay-Lussac's law apply to the normal operation of a forced-air type of air-conditioning system. How does the system move heat and produce a cooling effect?

_____

_____

_____

_____

_____

_____

_____

29. If air became trapped inside an air conditioner's condenser, how would this affect the condenser's pressure and temperature? Explain why and include which gas laws are at work.

_____

_____

_____

_____

_____

_____

_____

_____

Name _____

30. How would an air conditioner's pressure and temperature values be affected if it was operating in ambient temperature of 70°F and that temperature rose to 100°F? Remember that a temperature difference is used to cause heat transfer between refrigerant and ambient air. Which gas laws would be in effect? Explain how they would work and how they would affect the system's operating values (high-side and low-side temperatures and pressures).

_____

_____

_____

_____

_____

_____

_____

_____

_____

_____

Notes

CHAPTER 6

Name _C. COSTELLO_

Date _9-5-17_ **Class** _Refrigeration_

Basic Refrigeration Systems

> Carefully study the chapter and then answer the following questions.

6.1 Compression Refrigeration Cycle

True   1. *True or False?* Refrigerants evaporate to release heat outside of the refrigerated space.

_C_   2. The evaporation of a fluid is an indication that _____ is being absorbed.
- A. acid
- B. air
- C. heat
- D. moisture

_D_   3. Volume is decreased inside the compressor, causing an increase in pressure and _____.
- A. condensation
- B. evaporation
- C. molecules
- D. temperature

_C_   4. *True or False?* The state of refrigerant that is capable of absorbing the most heat is _____.
- A. high-pressure liquid
- B. high-pressure vapor
- C. low-pressure liquid
- D. low-pressure vapor

6.2 High Side and Low Side

False   5. *True or False?* The high side of the refrigeration system is where heat is absorbed and removed from the refrigerated space.

_D_   6. The high side of a refrigeration system is divided from the low side by the _____.
- A. liquid receiver
- B. condenser
- C. evaporator
- D. metering device

D 7. Low-side pressure may also be called _____ pressure.
   A. condenser
   B. discharge
   C. head
   D. suction /evaporator

6.3 Compression

C 8. The reciprocating movement of a compressor piston _____.
   A. provides suction
   B. compresses the refrigerant
   C. Both A and B.
   D. Neither A nor B.

D 9. Operational balance is reached when the number of vapor molecules that condense into liquid in the condenser equals the number of _____.
   A. liquid refrigerant molecules in the liquid receiver
   B. lubricant molecules returned to the crankcase
   C. liquid molecules needed to fill the evaporator
   D. vapor refrigerant molecules pumped by the compressor

A 10. An oil separator uses _____, or screens, to separate oil from refrigerant vapors.
   A. baffles
   B. magnets
   C. pumps
   D. vacuums

6.4 Condensing

C 11. A condenser's primary function in a system is to remove _____ heat from refrigerant.
   A. ambient
   B. compressor
   C. latent
   D. sensible

D 12. Which type of water-cooled condenser consists of only two tubes with water flowing in one direction in one tube and refrigerant flowing in the opposite direction in the other tube?
   A. Forced-water
   B. Shell-and-coil
   C. Shell-and-tube
   D. Tube-within-a-tube

Name _____

2

_____B_____ 13. The quantity of refrigerant in a system is less critical when the system has a(n) _____.

    A. liquid line solenoid valve
    B. liquid receiver
    C. oversized compressor
    D. refrigerant leak

_____C_____ 14. The _____ line connects the liquid receiver to the metering device.

    A. condensate
    B. discharge
    C. liquid
    D. suction

_____FALSE_____ 15. *True or False?* A liquid line filter-drier can also be called a low-side filter-drier.

6.5 Metering Device

_____B_____ 16. A metering device is intended to provide _____ in liquid refrigerant's passage into the evaporator.

    A. filtration
    B. a restriction
    C. superheat
    D. All of the above.

_____C_____ 17. The drop in pressure between the metering device and the evaporator _____ the boiling point of the refrigerant.

    A. does not change
    B. doubles
    C. lowers
    D. quadruples

6.6 Evaporating

_____B_____ 18. To create a cooling effect, an evaporator primarily absorbs _____ heat.

    A. adiabatic
    B. latent
    C. liquid
    D. sensible

_____False_____ 19. *True or False?* The term *superheated* means that the refrigerant is a liquid with a temperature below the boiling point.

_____False_____ 20. *True or False?* The purpose of a suction line filter-drier is to increase the pressure difference between the evaporator and the compressor.

Match the letter corresponding to the name of the refrigeration system component to its description below. Not all letters will be used.

_____D_____ 21. The liquid refrigerant entering this component is under low pressure, which makes the liquid vaporize and absorb heat.

_____I_____ 22. A device that removes oil from hot, compressed vapor.

_____J_____ 23. The line carrying refrigerant from the evaporator to the compressor.

_____C_____ 24. In this component, heat is removed from the refrigerant vapor.

_____E_____ 25. This component carries refrigerant from the condenser to the evaporator.

_____A_____ 26. A safety device to prevent liquid refrigerant from entering the suction line and compressor.

_____F_____ 27. This component prevents moisture, dirt, and metal chips from entering the metering device.

_____H_____ 28. This component controls liquid refrigerant flow into the evaporator.

A. Accumulator

B. Compressor

C. Condenser

D. Evaporator

E. Liquid line

F. Liquid line filter-drier

G. Liquid receiver

H. Metering device

I. Oil separator

J. Suction line

Critical Thinking

29. An earlier question asked which state of refrigerant is capable of absorbing the most amount of heat among the following four options: high-pressure liquid, high-pressure vapor, low-pressure liquid, and low-pressure vapor. Select the correct answer and explain why refrigerant in that state can absorb the most amount of heat (in Btu).

Low Pressure Liquid is like a Dry Sponge it can absorb sensible heat until it Boils then absorbing latent heat

30. Explain at least three ways that heat can enter a refrigerator.

Door is opened, crack in Seal, Suction line is exposed

Name _____

31. Explain the two functions of a compressor as they relate to the low side and high side of an HVAC system.

Low side - Compressor sucks from Evaperator
High Side - compressor forces Hot vapor thru condenser
Acts as heart

_____

_____

32. Explain why the discharge line is smaller than the suction line. What does this size difference indicate about the condition of the refrigerant within these two different lines? Is anything else different about the refrigerant in each line?

It is smaller because the Discharge is a liquid line
and liquid has Been condensed from a gas

_____

33. Name the two types of air-cooled condensers and explain their advantages and disadvantages.

Natural convection condenser - No Fan
Forced air condensor - has fan

_____

34. Explain how a metering device reduces the pressure of liquid refrigerant in the evaporator.

It regulates flow   so that refrigerant can flash
boil and expand in Tube. The Suction from
Compressor makes a lower pressure

_____

35. Explain how an accumulator prevents liquid refrigerant from reaching the compressor.

It traps liquid and holds it unit it Boils and
then releases it Back To suction line

_____

_____

Notes

CHAPTER 7

Tools and Supplies

Name _____

Date _____ Class _____

3

Carefully study the chapter and then answer the following questions.

7.1 Hand Tools

_____ 1. If a bolt is sized 1/2″ or larger, then its corresponding wrench size should be _____ larger than the bolt size.
   A. 1/8″
   B. 3/16″
   C. 1/4″
   D. 5/16″

_____ 2. If a bolt is sized 7/16″ or smaller, then its corresponding wrench size should be _____ larger than the bolt size.
   A. 1/8″
   B. 3/16″
   C. 1/4″
   D. 5/16″

Add numbers (from 1 through 4) to the four major types of wrenches in the order of preferred use. Number 1 is the most preferred, and 4 is the least preferred.

_____ 3. Adjustable wrench

_____ 4. Box end wrench

_____ 5. Open end wrench

_____ 6. Socket wrench

_____ 7. *True or False?* A nut driver is primarily used to turn screws.

_____ 8. For odd-size nuts and bolts, a(n) _____ wrench with movable jaws may be necessary.
   A. adjustable
   B. box end
   C. flare nut
   D. open end

_____ 9. To tighten or loosen fasteners with a star-shaped indent in the head, a(n) _____ wrench is used.
   A. Allen
   B. hex key
   C. service valve
   D. Torx®

_____ 10. What type of wrench is shown in the following image?
   A. Adjustable wrench
   B. Box end wrench
   C. Pipe wrench
   D. Service valve wrench

Reed Manufacturing Co.

_____ 11. The slight opening of a valve to cause the valve needle to leave its seat is called _____.
   A. cracking
   B. drifting
   C. flaring
   D. torquing

_____ 12. When striking parts to safely drive them together or separate them, it is best to use a _____.
   A. chisel
   B. claw hammer
   C. mallet
   D. screwdriver

_____ 13. *True or False?* Lineman's pliers can be used for both cutting and gripping.

| *Identify the types of specialty pliers shown in the following image.* |

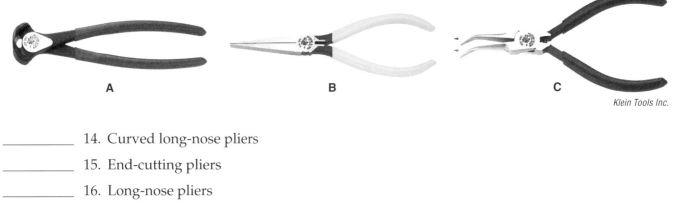

A   B   C

Klein Tools Inc.

_____ 14. Curved long-nose pliers

_____ 15. End-cutting pliers

_____ 16. Long-nose pliers

Name _____

_____ 17. *True or False?* It is a good idea to use combination (split-joint) pliers to loosen and tighten nuts and bolts, since it means not having to carry any wrenches.

_____ 18. *True or False?* When clamping parts that should not be marred, jaw pad inserts should be used on a vise so as not to damage the metal.

_____ 19. In general, the _____ a drill bit is, the faster it should be turned.
   A. larger
   B. longer
   C. shorter
   D. smaller

_____ 20. Always wear _____ when working with chisels or punches.
   A. ear protection
   B. eye protection
   C. a hard hat
   D. steel-toed boots

_____ 21. A tap drill should be slightly _____ than the inside diameter of the threads for which the hole is being drilled.
   A. larger
   B. longer
   C. smaller
   D. None of the above.

_____ 22. When punching the location of a hole to be drilled, it is best to use a _____ punch.
   A. center
   B. drift
   C. pin
   D. prick

Select the appropriate answer from the list to match the following statements. Not all letters will be used.

_____ 23. Easier to use than a box end wrench in restricted spaces.

_____ 24. Designed to grip cylindrical surfaces.

_____ 25. Single or double cut.

_____ 26. The correct way to apply pressure when using a wrench.

_____ 27. Number of points on a hex key wrench.

_____ 28. Measured by a torque wrench.

_____ 29. Maximum amount of strokes per minute that a hacksaw should be stroked.

_____ 30. Type of file used for finishing surfaces.

_____ 31. Used to set a line level, plumb, or at 45°.

A. 60

B. Amount of tightness

C. 8

D. Files

E. Level

F. Open end wrench

G. Pipe wrench

H. Pull

I. Push

J. Single cut

K. Six

7.2 Power Tools

_____ 32. When drilling into a hard surface, such as concrete or other masonry, it is best to use a(n) _____.
   A. electric drill
   B. hammer drill
   C. impact driver
   D. reciprocating saw

_____ 33. When quickly assembling and disassembling units, it would be best to use a(n) _____.
   A. data logger
   B. hammer drill
   C. impact driver
   D. reciprocating saw

_____ 34. When cutting holes through floors and walls for rectangular ductwork, air registers, and grilles, it would be best to use a(n) _____.
   A. electric drill
   B. hammer drill
   C. impact driver
   D. reciprocating saw

7.3 Instruments

_____ 35. Which type of thermometer can be used to record temperatures over a 7-day period?
   A. Analog
   B. Data logger
   C. Glass-stem
   D. Mercury

_____ 36. The basic use for a manometer in HVACR work is to measure _____.
   A. air velocity in ductwork
   B. head pressure
   C. manos
   D. suction pressure

_____ 37. What type of manometer is shown in the following image?
   A. Digital manometer
   B. Inclined manometer
   C. Rankine manometer
   D. U-tube manometer

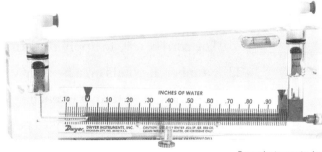

Dwyer Instruments, Inc.

Name _____

7.4 Standard Supplies

_____ 38. *True or False?* Machine screws are commonly used when parts are to be put together permanently.

_____ 39. Which of the following is *not* true of gaskets?
  A. Gaskets conform to irregularities in mating surfaces.
  B. Gaskets are commonly used between the valve plate and the compressor body.
  C. Replacement gaskets should be at least 1/2" thicker than the original gaskets.
  D. Gaskets must not restrict any passageway openings.

_____ 40. Which of the following can be used to clean refrigeration mechanisms?
  A. Carbon tetrachloride
  B. Gasoline
  C. Oleum (mineral spirits)
  D. Propane

7.5 Employer-Provided Tools and Equipment

_____ 41. Employer-provided tools and equipment generally include the following, *except* _____.
  A. hand tools
  B. recovery/recycling unit
  C. soldering, brazing, and welding outfit
  D. vacuum pump

Critical Thinking

42. Explain how to hold a hammer to deliver light, accurate blows.

_____

_____

_____

_____

_____

43. Explain how to stroke a hacksaw to help keep the cutting edge sharp. Explain why this method keeps the teeth sharp.

_____

_____

_____

_____

_____

44. List three commonly used abrasives other than sandpaper and describe some applications for their use. Why would one abrasive work better in one situation than another abrasive?

_____

_____

_____

_____

_____

_____

_____

45. Why do you think that service companies provide some tools and instruments for technicians but not other tools and instruments?

_____

_____

_____

_____

_____

_____

CHAPTER 8

Working with Tubing and Piping

Name _____

Date _____ Class _____

Carefully study the chapter and then answer the following questions.

8.1 Types of Refrigerant Tubing and Pipe

_____ 1. The most common type of tubing used in refrigeration systems is _____.
   A. aluminum
   B. copper
   C. plastic
   D. steel

_____ 2. ACR tubing that is Type L is _____ wall.
   A. double
   B. heavy
   C. medium
   D. thin

_____ 3. ACR tubing that is Type K is _____ wall.
   A. double
   B. heavy
   C. medium
   D. thin

_____ 4. *True or False?* To speed up the annealing process, pour water on tubing that has just been heated to cool it more quickly.

_____ 5. *True or False?* Hard-drawn ACR tubing must be connected by brazing, because hard-drawn tubing cannot be flared.

_____ 6. A type of strong tubing that is resistant to corrosion and commonly used for food processing, milk handling, and other refrigeration applications is _____ tubing.
   A. aluminum
   B. cadmium
   C. PEX
   D. stainless steel

_____ 7. A type of tubing commonly used to form evaporators and is less costly than copper is _____ tubing.
   A. aluminum
   B. cadmium
   C. PEX
   D. stainless steel

8.2 Non-Refrigerant Tubing and Pipe

_____ 8. *True or False?* Copper water tubing is basically the same as copper ACR tubing, so technicians can safely use either tubing interchangeably for either purpose.

_____ 9. The nominal size of copper water tubing is _____ than its outside diameter.
   A. 1/8″ less
   B. 1/8″ more
   C. 1/4″ less
   D. 1/4″ more

_____ 10. Plastic tubing used in refrigeration-related work includes the following, *except* _____.
   A. ABS
   B. ACR
   C. CPVC
   D. PVC

_____ 11. *True or False?* CPVC pipe has a maximum service temperature higher than that of PVC pipe.

_____ 12. Black steel pipe in refrigeration-related work is generally used for conveying _____.
   A. air
   B. gas
   C. refrigerant
   D. water

8.3 Cutting Tubing

_____ 13. Tubing should always be cut at an angle of _____.
   A. 30°
   B. 45°
   C. 75°
   D. 90°

_____ 14. After tubing is cut, burrs on the cut ends must be removed by _____.
   A. annealing
   B. fluxing
   C. reaming
   D. swaging

Name _____

8.4 Bending Tubing

_____ 15. The minimum radius for a tubing bend is _____ times the diameter of the tubing.

   A. 2
   B. 5
   C. 10
   D. 20

_____ 16. When using an external spring to bend soft ACR tubing that will be connected to a flare connector, remember to _____.

   A. anneal the tubing before bending
   B. bend the tubing before flaring
   C. flare the tubing before bending
   D. All of the above.

8.5 Connecting Tubing

_____ 17. A standard single flare on copper tubing is made at a _____ angle.

   A. 15°
   B. 20°
   C. 37°
   D. 45°

_____ 18. If a flare fitting connects 3/8″ tubing to 7/16″ tubing, it is called a _____ reducing fitting.

   A. 3/8″ to 7/16″
   B. 7/16″ to 3/8″
   C. Both A and B.
   D. Neither A nor B.

_____ 19. *True or False?* Soldering is commonly used to connect tubing that carries refrigerant.

_____ 20. For the hottest flame possible, which should you use?

   A. An air-acetylene torch
   B. A butane torch
   C. An oxyacetylene torch
   D. A propane torch

_____ 21. The type of torch flame that is the most efficient to use when brazing is the _____ flame.

   A. carburizing
   B. neutral
   C. oxidizing
   D. All of the above.

_____ 22. The movement of a liquid substance between two solid substances due to the adhesive forces of the solids overcoming the liquid's cohesive forces describes _____.

   A. annealing
   B. capillary action
   C. carburization
   D. flashback arresting

_____ 23. The paste or powder that prevents oxide from forming and aids solder flow is called _____.
   A. epoxy
   B. flux
   C. lubricant
   D. solvent

_____ 24. *True or False?* If solder does *not* melt on contact with the heated joint, use the torch to melt the solder directly.

_____ 25. With a horizontal joint, begin soldering by touching the solder to the _____ of the joint.
   A. bottom
   B. side
   C. top
   D. All of the above.

_____ 26. Which of the following can be used safely to purge tubing before brazing?
   A. Compressed air
   B. Nitrogen
   C. Oxygen
   D. Refrigerant

_____ 27. *True or False?* The strength of a brazed joint depends more on the type of filler metal used than on the clearance between the tubing and the fitting.

Flux behavior is a good indication of the temperature of a joint being heated for brazing. As the temperature of flux rises, its behavior changes in a certain sequence. Match the letter of the proper temperature level to the behavior of flux during the brazing process.

_____ 28. Flux lies flat and has a milk appearance.   A. 212°F (100°C)

_____ 29. Flux turns into a clear liquid.   B. 600°F (316°C)

_____ 30. Flux bubbles and remains white.   C. 800°F (427°C)

_____ 31. Flux turns white and somewhat puffy.   D. 1100°F (593°C)

_____ 32. Which of the following is *not* a type of tool used to swage tubing?
   A. Hydraulic swaging tool
   B. Lever swaging tool
   C. Swaging blade
   D. Swaging punch

_____ 33. *True or False?* Mechanical fittings can be used to join two tubes made of dissimilar metals.

8.6 Connecting Pipe

_____ 34. A pipe's strength is based on its wall thickness, which is called pipe _____.
   A. barreling
   B. bore
   C. capacity
   D. schedule

Name _____

_____ 35. The taper rate for tapered pipe threads is 1 unit of diameter for every _____ unit(s) of
length.
   A. 1
   B. 2
   C. 6
   D. 16

_____ 36. A street fitting is an angled tube or pipe fitting with _____.
   A. one male end and one female end
   B. a threaded mechanical joint in its center
   C. two female ends
   D. two male ends

_____ 37. Two plastic pipes can be joined through a process called _____ welding, which allows the
surface polymers of each plastic pipe to dissolve and become entangled.
   A. flange
   B. flare
   C. solvent
   D. swage

_____ 38. *True or False?* CPVC cement should be used to join Schedule 40 PVC pipe.

Critical Thinking

39. Explain the reasons why it is preferable to cut ACR tubing using a wheel-type tubing cutter rather than
a hacksaw. What could happen to an HVACR system using tubing cut with a hacksaw?

_____

_____

_____

_____

_____

_____

_____

40. Why is a large radius preferable when making bends in soft ACR tubing?

_____

_____

_____

_____

_____

41. Why should the flaring cone on a flaring tool be slowly tightened and then backed off repeatedly to form a flare?

_____

_____

_____

42. Why would an oxyacetylene torch be more effective at brazing joints than an air-acetylene torch?

_____

_____

_____

43. Why should brazing filler rod be applied to a joint at a 30° to 45° angle?

_____

_____

_____

44. When running piping and dealing with an offset or other configuration, explain as many benefits as possible of using street fittings instead of standard fittings.

_____

_____

_____

_____

_____

_____

_____

_____

CHAPTER 9

Introduction to Refrigerants

Name _____

Date _____ Class _____

| Carefully study the chapter and then answer the following questions. |

9.1 Refrigerants and the Ozone Layer

_____ 1. The US governmental body charged with enforcing the regulations of the Clean Air Act is the _____.

    A. ASHRAE

    B. DOT

    C. EPA

    D. SNAP

_____ 2. Section _____ of the Clean Air Act requires that all technicians have certification in order to service HVACR equipment.

    A. 8

    B. 9

    C. 179

    D. 608

_____ 3. Which type of refrigerant has the highest ozone depletion potential?

    A. CFCs

    B. HCFCs

    C. HFCs

    D. HFOs

_____ 4. HFCs have an ozone depletion potential of _____.

    A. 0

    B. 1.5

    C. 11

    D. 22

_____ 5. A refrigerant's global warming potential (GWP) is based on the ratio of its warming effect compared to the warming effect of _____.

    A. ammonia

    B. carbon dioxide

    C. oxygen

    D. R-11

9.2 Classifying Refrigerants

_____   6. A partially halogenated refrigerant classification with 0 ODP but high GWP describes _____.
   A. HCs
   B. HCFCs
   C. HFCs
   D. HFOs

_____   7. A refrigerant classification with 0 ODP, low GWP, low toxicity, slightly flammable, and used in some automatic air-conditioning systems describes _____.
   A. HCs
   B. HCFCs
   C. HFCs
   D. HFOs

_____   8. A halogenated refrigerant classification with a high ODP and no longer manufactured since 1995 describes _____.
   A. CFCs
   B. HCFCs
   C. HFCs
   D. HFOs

_____   9. A refrigerant classification of organic substances with 0 ODP, low GWP, low toxicity, but high flammability and used in standalone freezers and domestic refrigerator-freezers describes _____.
   A. HCs
   B. HCFCs
   C. HFCs
   D. HFOs

_____   10. A refrigerant classification with a moderate ODP level but high GWP with a phaseout scheduled for 2030 describes _____.
   A. CFCs
   B. HCs
   C. HCFCs
   D. HFCs

_____   11. If a refrigerant blend contains an HCFC and an HFC, it is considered a(n) _____.
   A. CFC
   B. HCFC
   C. HFC
   D. HFO

_____   12. Which refrigerant blend responds to temperature change like a single refrigerant?
   A. azeotropic
   B. near-azeotropic
   C. zeotropic
   D. None of the above.

Name _____

_____ 13. The separating of a zeotropic blend's individual refrigerants during phase change is called _____.

    A. carbonation
    B. flashing
    C. floccing
    D. fractionation

9.3 Identifying Refrigerants

_____ 14. Often, the number assigned to a specific refrigerant is associated with the refrigerant's _____.

    A. chemistry
    B. GWP
    C. popularity
    D. price

Match the letter of the proper cylinder color corresponding to each refrigerant listed.

_____ 15. R-12

_____ 16. R-22

_____ 17. R-134a

_____ 18. R-404A

_____ 19. R-410A

_____ 20. R-508B

A. dark blue

B. light blue (sky)

C. light green

D. orange

E. pink (rose)

F. white

9.4 Refrigerant Properties

_____ 21. A refrigerant that is classified as A1 has _____.

    A. high toxicity and high flammability
    B. high toxicity and no flammability
    C. low toxicity and high flammability
    D. low toxicity and no flammability identified

_____ 22. A refrigerant that is classified as B3 has _____.

    A. high toxicity and high flammability
    B. high toxicity and no flammability
    C. low toxicity and high flammability
    D. low toxicity and no flammability identified

_____ 23. When measuring low-side pressure, a technician can use a pressure-temperature (P/T) chart to find ____.
  A. the condenser's subcooling value
  B. the evaporator's superheat value
  C. high-side refrigerant's condensing temperature
  D. low-side refrigerant's boiling temperature

_____ 24. The temperature of tubing in a refrigeration system is referred to as ____ temperature.
  A. ambient
  B. conditioned space
  C. drop
  D. skin

_____ 25. When the compressor is running, the temperature of the evaporator tubing is about ____ than the temperature of the refrigerant within.
  A. 10°F cooler
  B. 10°F warmer
  C. 20°F cooler
  D. 20°F warmer

_____ 26. When the compressor is running on a system with a standard air-cooled condenser, the temperature difference between ambient air and condenser refrigerant is about ____.
  A. 5°F
  B. 10°F
  C. 20°F
  D. 30°F

9.5 Refrigerant Applications

_____ 27. *True or False?* Even though some refrigerants, such as R-12, are illegal to manufacture, it is still legal to service units containing R-12.

Match each of the following refrigerants with its common use.

_____ 28. HFO refrigerant used in automotive air conditioning.

_____ 29. Popular HFC refrigerant used in domestic and commercial refrigeration and as a retrofit for R-12.

_____ 30. Zeotropic blend used to retrofit R-22 equipment.

_____ 31. Azeotropic blend used in low-temperature refrigeration.

_____ 32. HFC zeotropic blend used in medium- and low-temperature refrigeration.

_____ 33. HFC zeoptropic blend used in new air-conditioning equipment.

_____ 34. HC refrigerant commonly used in domestic refrigerators, freezers, and standalone equipment.

A. R-134a
B. R-404A
C. R-407C
D. R-410A
E. R-508B
F. R-600a
G. R-1234yf

Name _____

9.6 Inorganic Refrigerants

_____ 35. Which of the following is *not* a characteristic of R-717?

A. It corrodes copper and bronze.

B. It is flammable at 150,000 ppm.

C. It is odorless.

D. It is used in absorption systems.

_____ 36. The most commonly used expendable refrigerants include the following, *except* _____.

A. ammonia (R-717)

B. carbon dioxide (R-744)

C. helium (R-704)

D. nitrogen (R-728)

_____ 37. Liquid refrigerants that are used to quickly freeze food products are called _____.

A. azeotropes

B. dew blends

C. freezants

D. halogens

9.7 Refrigeration Lubricants

_____ 38. A refrigeration lubricant with a high _____ point may have excessive wax separation, which can clog the system.

A. bubble

B. dew

C. flash

D. floc

_____ 39. A measure of a liquid's resistance to flow is called _____.

A. dielectric strength

B. pour point

C. thermal stability

D. viscosity

_____ 40. Regarding pour point, a refrigeration lubricant should be selected based on a system's _____.

A. ambient temperature

B. operating temperatures

C. subcooling

D. superheat

_____ 41. A group of synthetic lubricants compatible with CFCs, HCFCs, and HFCs describes _____ lubricants.

A. AB

B. MO

C. PAG

D. POE

_____ 42. A refrigeration lubricant made from refined crude oil and used only with CFCs and HCFCs describes _____ lubricants.

    A. AB

    B. MO

    C. PAG

    D. POE

_____ 43. Lubricants specifically designed for use with HFCs and are not recommended for use in compressors with aluminum pistons in steel cylinders describes _____ lubricants.

    A. AB

    B. MO

    C. PAG

    D. POE

_____ 44. Lubricants made from propylene and benzene and designed for use with CFCs, HCFCs, and blends that include CFCs and HCFCs describes _____ lubricants.

    A. AB

    B. MO

    C. PAG

    D. POE

_____ 45. _True or False?_ Refrigeration lubricants are stable fluids that should be kept in open containers or cylinders that are vented to atmosphere.

_____ 46. When refrigeration lubricant is found to be dark black in color, pungent in odor, and containing metal shavings, it is likely that the _____ failed.

    A. compressor

    B. condenser

    C. evaporator

    D. refrigerant metering device

Critical Thinking

47. What is temperature glide? How does temperature glide differentiate the types of refrigerants? In your own words, explain what is happening in a refrigeration system during temperature glide.

_____

_____

_____

_____

_____

_____

_____

Name _____

> *Use the values provided below to determine the refrigeration effect, heat of compression, and coefficient of performance of a refrigeration system that uses a refrigerant with the following enthalpies listed. Show your calculations and work for each answer.*

- Heat content of refrigerant entering the evaporator: 49 Btu/lb
- Heat content of refrigerant leaving the evaporator: 110 Btu/lb
- Heat content of refrigerant entering the condenser: 127 Btu/lb

_____ 48. Refrigeration effect

_____ 49. Heat of compression

_____ 50. Coefficient of performance

51. List at least four properties that new refrigerants should have. In your own words, explain why each of these traits is desirable and beneficial. What might happen if a refrigerant did not have one of these traits?

_____

_____

_____

_____

_____

_____

52. What is the problem with opening a container of refrigeration lubricant a few hours before adding it to a system? What could happen?

_____

_____

_____

_____

_____

_____

CHAPTER 10

Name _C. COSTELLO_

Date _9-23-17_ **Class** _____

Equipment and Instruments for Refrigerant Handling and Service

> *Carefully study the chapter and then answer the following questions.*

10.1 Refrigerant Cylinders

_____A_____ 1. Cylinders that contain a corrosive refrigerant must be inspected every _____ years.
  A. 1
  B. 5
  C. 10
  D. 25

_____B_____ 2. Cylinders that contain a noncorrosive refrigerant must be inspected every _____ years.
  A. 1
  B. 5
  C. 10
  D. 25

_____D_____ 3. At a service shop, a refillable service cylinder would be topped off using a _____ cylinder.
  A. disposable
  B. nitrogen
  C. recovery
  D. storage

_____B_____ 4. Disposable cylinders should be stored at temperatures below _____ to prevent refrigerant pressure buildup.
  A. 55°F (13°C)
  B. 125°F (51°C)
  C. 150°F (66°C)
  D. 175°F (80°C)

_____B_____ 5. Which two colors are recovery cylinders painted in order to distinguish them from disposable cylinders?
  A. Blue and gray
  B. Gray and yellow
  C. Orange and gray
  D. Yellow and blue

_____ C   6. A disposable cylinder should be evacuated down to _____ before being disposed of or recycled.
   A. 15 in. Hg
   B. 15 kPa
   C. 15 psia
   D. 15 psig

_____ C   7. When charging as much liquid refrigerant as possible into a system from a disposable cylinder, the cylinder should be placed _____.
   A. on its side
   B. right side up
   C. upside down
   D. None of the above.

_____ B   8. Recovery cylinders have two valves, which are generally marked _____.
   A. in and out
   B. liquid and gas
   C. open and close
   D. vapor and vent

10.2 Pressure Gauges

_____ A   9. The device inside a mechanical pressure gauge that straightens as pressure is increased is called a(n) _____.
   A. Bourdon tube
   B. retaining rod
   C. retarder
   D. thermocouple

_____ C   10. Which of the following measurements is the equivalent of atmospheric pressure at sea level?
   A. 1 torr
   B. 25.4 mm Hg
   C. 29.92 in. Hg
   D. 25,400 microns

_____ FALSE   11. _True or False?_ A compound gauge is intended to be used primarily on the high side of a system.

10.3 Service Valves

_____ A   12. When a system is running in normal operation, a service valve should be _____.
   A. back seated
   B. cracked open
   C. front seated
   D. in mid-position

Name _____C. COSTELLO_____

_____C_____ 13. When a service valve blocks the flow of refrigerant through the valve by closing off its regular passageway, it is _____.

A. back seated
B. cracked open
C. front seated
D. in mid-position

_____D_____ 14. When a service valve stem is turned so that it is halfway between front and back seats, it is considered _____.

A. back seated
B. cracked open
C. front seated
D. in mid-position

_____B_____ 15. When a service valve stem is turned so that it is just lifted off the back seat, it is considered _____.

A. back seated
B. cracked open
C. front seated
D. in mid-position

_____A_____ 16. Which of the following is a compressor service valve on the high side of a system?

A. Discharge service valve
B. Liquid receiver service valve
C. Suction line service valve
D. Suction service valve

_____ 17. To evacuate a system more quickly through an access port, remove the _____ from a Schrader valve.

A. fusible plug
B. retarder
C. threads
D. valve core

_____B_____ 18. A refrigeration system should be _____ before a brazed-on piercing valve is installed.

A. completely disassembled
B. recovered and evacuated
C. retrofitted with new refrigerant
D. sold for scrap metal

10.4 Gauge Manifolds

Match the following components of a gauge manifold with the corresponding letters in the image.

___B___ 19. High-side valve

___H___ 20. Low-side port

___A___ 21. High-pressure gauge

___E___ 22. Connection to refrigerant cylinder, recovery machine, or vacuum pump

___J___ 23. Compound gauge

___C___ 24. High-side port

___G___ 25. Sight glass

Imperial

___C___ 26. To reduce the amount of refrigerant released to the atmosphere when connecting hoses, _____ fittings can be used.
   A. fusible plug
   B. free air displacement
   C. quick-connect
   D. retarder

___B___ 27. Which of the following corresponds to the positions of a gauge manifold's valves during system evacuation?
   A. Both valves closed
   B. Both valves open
   C. High-side valve open, low-side valve closed
   D. Low-side valve open, high-side valve closed

___False___ 28. *True or False?* Gauges used at locations of high elevations should be calibrated to 0 psig at sea level.

Name _C. COSTELLO_

_____C____ 29. Which of the following is illustrated in the following image?
  A. Evacuation
  B. Liquid charging
  C. Vapor charging
  D. All of the above.

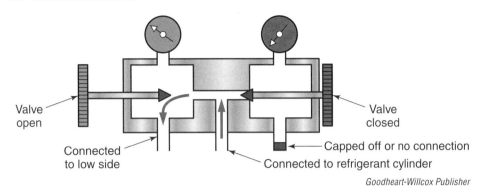

Valve open

Valve closed

Connected to low side

Capped off or no connection

Connected to refrigerant cylinder

Goodheart-Willcox Publisher

10.5 Leak Detection Devices

Match the following types of leak detection devices with the phrase that best describes them.

_____D_____ 30. Uses a rubber hose to "sniff" for leaks.

_____C_____ 31. Requires an ultraviolet light to scan for leaks.

_____A_____ 32. A low-cost method used to confirm exact leak locations.

_____E_____ 33. Detects the sound of refrigerant escaping under pressure.

_____B_____ 34. May require a plastic tip guard so the probe tip does not contact surfaces.

A. Bubble solution
B. Electronic detector
C. Fluorescent dye
D. Halide torch
E. Ultrasonic detector

_____True_____ 35. _True or False?_ Electronic leak detectors may falsely indicate a leak if they are used around urethane insulation.

10.6 Vacuum Pumps

_____ 36. Most manufacturers recommend pulling a vacuum pressure of _____ prior to recharging a system.
  A. 500 in. Hg
  B. 500 microns
  C. 500 mm Hg
  D. 500 torr

Match the following HVACR system size with the proper corresponding vacuum pump capacities.

_____ 37. Over 100 tons

_____ 38. 3- to 5-ton residential systems

_____ 39. 5- to 100-ton medium systems

A. 1.5 cfm

B. 3–5 cfm

C. 10–15 cfm

_____ 40. The purpose of oil in a vacuum pump is to act as a lubricant for the pump and also as a(n) _____.
   A. aid in pressure measurement
   B. fluid seal between air, gases, and contaminants entering the pump
   C. system leak detection method
   D. All of the above.

10.7 Recovery, Recycling, and Reclaiming Equipment

_____ 41. Recovery/recycling machines must be certified to comply with AHRI Standard _____.
   A. 210/240
   B. 340-360
   C. 410
   D. 740

_____ 42. *True or False?* Recovered refrigerant can be sold to a new owner once it has been recycled.

_____ 43. Before repairing a leak, all of the refrigerant in a system must be _____.
   A. recovered
   B. recycled
   C. reclaimed
   D. resold

_____ 44. If a refrigerant cylinder weighs 100 lb and its tare weight is 20 lb, how much refrigerant was removed if the final weight is 50 lb after charging a system?
   A. 20 lb
   B. 30 lb
   C. 50 lb
   D. 80 lb

Critical Thinking

45. After topping off an air-conditioning system that had a low refrigerant charge, you close the cylinder's hand valve. Explain what safety procedure steps should be followed when moving this cylinder back to your service truck. Note that this refrigerant cylinder weighs over 35 lb.

_____

_____

_____

_____

Name _____

46. An HVAC service shop often has large storage cylinders of frequently used refrigerant available. These service cylinders often have valves on the bottom or are positioned upside down so the valve appears on the bottom. What are the advantages and reasons for doing this?

_____

_____

_____

_____

_____

4

47. An HVAC technician goes on an emergency service call on a cool night with outside air at 47°F (8.3°C). The customer's heat pump is malfunctioning, and the technician decides to turn on the emergency heat function, recover the entire refrigerant charge, pull a vacuum overnight, and troubleshoot the system malfunction in the morning. The only available recovery cylinder was already 70% full of liquid. After the refrigerant recovery from the heat pump, the recovery cylinder is 95% full of liquid. That night the technician writes a brief report and decides to take the next day off. The next morning, the paperwork gets misplaced, and the 95% full recovery cylinder is left sitting in the back of a service truck where it is exposed to full sun and ambient temperature rises to 120°F (49°C) by 9:30 am with no indication of dropping. Explain what might happen to the refrigerant in the recovery cylinder and include any reasons why.

_____

_____

_____

_____

_____

_____

_____

_____

48. List the three main types of pressure gauges used in HVACR work and explain how they are commonly used.

_____

_____

_____

_____

_____

49. List three things that a technician can do to maintain service valves and extend their service life.

_____

_____

_____

_____

50. Which service valve position helps prevent damage to gauges when pressures are being measured during system operation? Explain why.

_____

_____

_____

51. Explain why an oily or dusty area on refrigeration equipment would indicate a possible leak.

_____

_____

_____

52. Why is it useful to block off a suspected leak area from light and wind before using certain refrigerant leak detection methods?

_____

_____

_____

_____

_____

53. What two vacuum pump specifications are used to evaluate pump performance? What does each specification measure?

_____

_____

_____

_____

_____

_____

_____

54. What will happen if dirty oil is left in a vacuum pump?

_____

_____

_____

_____

Name _____

55. List at least four maintenance procedures that will extend the life of a recovery machine.

_____

_____

_____

_____

_____

56. Explain the difference between recycling and reclaiming refrigerant.

_____

_____

_____

_____

_____

_____

_____

_____

_____

Notes

Working with Refrigerants

Carefully study the chapter and then answer the following questions.

11.1 Checking Refrigerant Charge

_____ 1. Checking a system's refrigerant charge is important because an improper charge can _____.

A. lead to compressor damage
B. reduce system capacity and efficiency
C. waste electrical energy
D. All of the above.

_____ 2. Checking refrigerant charge by superheat is normally done on systems using _____.

A. any metering device
B. fixed orifice metering devices
C. thermostatic expansion valves
D. None of the above.

_____ 3. Checking refrigerant charge by subcooling is normally done on systems using _____.

A. any metering device
B. fixed orifice metering devices
C. thermostatic expansion valves
D. None of the above.

_____ 4. Subcooling may be described as the amount of heat _____.

A. added to refrigerant after it has evaporated
B. expelled from refrigerant passing through a condenser
C. removed from a refrigerant after it has condensed
D. removed from conditioned air passing through an evaporator coil

_____ 5. Superheat may be described as the amount of heat _____.

A. added to refrigerant after it has evaporated
B. expelled from refrigerant passing through a condenser
C. removed from a refrigerant after it has condensed
D. removed from conditioned air passing through an evaporator coil

_____ 6. What is the superheat of a system with a suction line temperature of 52°F, low-side pressure corresponding to 34°F, and a liquid line temperature of 100°F?

    A. 18°F

    B. 48°F

    C. 66°F

    D. 86°F

11.2 Redistributing Refrigerant

_____ 7. The first valve to front seat when performing a pumpdown is the _____.

    A. discharge service valve

    B. king valve

    C. queen valve

    D. suction service valve

_____ 8. _True or False?_ During passive recovery, raising the temperature of a recovery cylinder causes more refrigerant to flow from the system into the cylinder.

_____ 9. Some refrigerant cylinders have a liquid level switch that turns off the recovery machine when the cylinder is _____ full.

    A. 10%

    B. 25%

    C. 50%

    D. 80%

_____ 10. To perform liquid recovery, a recovery machine must be equipped with a(n) _____ that flashes liquid refrigerant into a vapor.

    A. check valve

    B. filter-drier

    C. restriction device

    D. sight glass

_____ 11. During push-pull liquid recovery, the recovery cylinder vapor port is connected to the _____.

    A. high-side service valve

    B. low-side service valve

    C. recovery machine inlet

    D. recovery machine outlet

_____ 12. To check whether a recovery cylinder's liquid level switch is working, connect the switch to a recovery machine and recovery cylinder setup, turn on the machine, and then _____.

    A. fill the recovery cylinder with water to over 80%

    B. raise the cylinder to high pressure

    C. recover the charge from the cylinder

    D. turn the recovery cylinder upside down

Name _____

11.3 Locating and Repairing Refrigerant Leaks

_____ 13. A system leak might be indicated by _____.

    A. a lack of cooling

    B. low suction and head pressures

    C. All of the above.

    D. None of the above.

_____ 14. If a system's recommended testing pressures are not known, the pressure should not exceed _____ when testing with carbon dioxide or nitrogen.

    A. 500 microns

    B. 14.7 psig

    C. 30 psig

    D. 170 psig

_____ 15. When brazing to repair a system leak, the nitrogen should be _____ in order to create a good brazed joint.

    A. high-pressure and flowing

    B. high-pressure and static

    C. low-pressure and flowing

    D. low-pressure and static

_____ 16. *True or False?* A two-part epoxy compound should only be purchased on an as-needed basis due to its limited shelf life.

11.4 Evacuating a System

_____ 17. A vacuum causes moisture in a system to evaporate by lowering the pressure in the system, which in turn lowers the _____ of any moisture in the system.

    A. balance point

    B. boiling point

    C. condensing point

    D. floc point

_____ 18. A refrigeration system is leak free if deep vacuum pressure holds steady around _____ or less.

    A. 25 microns

    B. 500 microns

    C. 1500 microns

    D. 2500 microns

_____ 19. To pull a deep vacuum as quickly as possible, what should a technician do regarding Schrader valves?

    A. Depress the pin to open the passageway.

    B. Pull out the Schrader pin.

    C. Remove the Schrader valve.

    D. Use a quick-connect fitting to prevent de minimus losses.

_____ 20. When pulling a vacuum, it is best to use hoses that are _____.

    A. long with large diameter

    B. long with small diameter

    C. short with large diameter

    D. short with small diameter

_____ 21. When pulling a vacuum, a technician can observe the gas laws in action. For example, in a refrigeration system, as pressure drops, _____.

    A. temperature drops

    B. temperature rises

    C. volume drops

    D. volume increases

_____ 22. If a vacuum must be pulled for a long time and must be left unattended by a technician, it is good practice to install a solenoid valve between the system and the vacuum pump. Such a solenoid valve should be wired _____ with the vacuum pump.

    A. in parallel

    B. in series

    C. Either A or B.

    D. Neither A nor B.

11.5 Charging a System

_____ 23. Which of the following is *not* a likely indication that a system may need to be charged with more refrigerant?

    A. High head pressure.

    B. Low low-side pressure.

    C. System short cycling.

    D. Visible leak (oil spots).

_____ 24. The quickest method of refrigerant charging is _____ charging.

    A. hand pump

    B. liquid

    C. push-pull

    D. vapor

_____ 25. Used for "topping off" a system, a(n) _____ cylinder holds a small amount of refrigerant, usually around 5 lb or less.

    A. charging

    B. recovery

    C. top-off

    D. vacuum

_____ 26. *True or False?* A professional technician always confirms the identity of a refrigerant by smelling and tasting it rather than relying on cylinder color alone.

_____ 27. When retrofitting a refrigeration system from one refrigerant to another, the following are likely to need replacing, *except* for the _____.

    A. filter-drier

    B. lubricant

    C. refrigerant metering device

    D. system thermostat

Name _____

Critical Thinking

28. Why is the subcooling of a system important? How does it affect both the low side of the system and the system overall?

_____

_____

_____

_____

_____

_____

_____

29. In order to check a system's superheat and subcooling, what tools, instruments, and supplies will be necessary?

_____

_____

_____

_____

_____

_____

30. When performing a system pump-down, which service valve is adjusted first? In what position should that valve be? Explain why that valve should be adjusted to that position.

_____

_____

_____

_____

_____

_____

_____

31. Your service manager sends you to relieve another technician whose shift is ending. The technician has already begun a system pump-down procedure on a large system with a liquid receiver, but you need to finish the process and perform service on the condenser. Which service valve—the discharge service valve, the king valve, or the queen valve—would you close to isolate the majority of the refrigerant charge in the smallest possible volume? Explain your answer.

_____

_____

_____

_____

_____

32. Why is passive recovery _not_ the best option for recovering refrigerant from a system that has a compressor that is not working?

_____

_____

_____

33. What is the advantage of using the liquid recovery method? Why must the vapor recovery method be used in combination with the liquid recovery method?

_____

_____

_____

_____

34. List four circumstances where the push-pull liquid recovery method _cannot_ be used.

_____

_____

_____

35. If a refrigeration system cannot hold a vacuum after being evacuated, then it probably has a leak. Explain why this is the case. What is causing the vacuum to fail?

_____

_____

_____

_____

Name _____

36. When testing for leaks, why is it preferable to charge a system with an inert gas instead of using more refrigerant?

_____

_____

_____

37. Evacuation is intended to remove all vapors and moisture from a system. As evacuation occurs, various measurable factors can be observed to show the different gas laws in action. A pressure drop that is too rapid can cause an evacuation problem. What is this problem and what steps can be taken to solve this problem?

_____

_____

_____

_____

_____

_____

38. Why are piercing valves impractical for use during deep-vacuum evacuation? Explain your reasoning.

_____

_____

_____

_____

_____

_____

39. Explain why a small amount of nitrogen is charged into a system after pulling the first and second vacuum for a triple evacuation process. What is the purpose of the nitrogen in this procedure?

_____

_____

_____

_____

_____

_____

40. Explain why a zeotropic refrigerant blend should be charged as a liquid. What might happen if a zeotrope was vapor charged instead?

_____

_____

_____

_____

41. How does R-410A service equipment differ from R-22 service equipment? Why does it differ?

_____

_____

_____

_____

CHAPTER 12

Basic Electricity

Name _____

Date _____ **Class** _____

5

Carefully study the chapter and then answer the following questions.

12.1 Fundamental Principles of Electronics

_____ 1. The nucleus of an atom is made of neutrons and _____.

    A. coulombs

    B. ions

    C. electrons

    D. protons

_____ 2. *True or False?* A positively charged ion is an atom that has more protons than electrons.

Match the following electrical terms with the phrase that best describes them. Not all letters will be used.

_____ 3. Electrical property that measures how much a material resists the flow of electrons.

_____ 4. A mathematical formula that explains the relationship among voltage, current, and resistance.

_____ 5. Unit used to measure electromotive force.

_____ 6. The potential difference in atomic charges that causes electricity to flow.

_____ 7. A measure of the electrical charge in 6.24×10^{18} electrons.

_____ 8. Unit used to measure the flow of current.

_____ 9. Equation used to calculate the resistance in a circuit.

_____ 10. The flow of electrons in a circuit.

A. Ampere

B. Coulomb

C. Current

D. $E \div I$

E. $E \div R$

F. Electromotive force

G. Ohm

H. Ohm's law

I. Resistance

J. Volt

_____ 11. According to Ohm's law, if the voltage in a circuit remains constant and resistance is increased, then _____.

    A. capacitance will increase
    B. coulombs will increase
    C. emf will decrease
    D. current will decrease

_____ 12. The conducting surfaces of a capacitor are separated by a _____.

    A. dielectric
    B. semiconductor
    C. terminal
    D. winding

_____ 13. Before removing or servicing a high-voltage capacitor in an HVACR system, discharge it with a _____ resistor.

    A. $20 \, \Delta\Omega$
    B. $20 \, k\Omega$
    C. $20 \, m\Omega$
    D. $20 \, \mu\Omega$

12.2 Types of Electricity

_____ 14. Static electricity is best understood as the _____.

    A. accumulation of an electric charge
    B. breaking of protons away from a nucleus
    C. flow of electrons
    D. varied dispersal of atoms in opposite directions

_____ 15. *True or False?* Direct current reverses the direction of electron flow in regular intervals.

_____ 16. Alternating current can be changed into direct current using a(n) _____.

    A. capacitor
    B. inverter
    C. rectifier
    D. transformer

_____ 17. In one cycle of alternating current, how many times does the maximum voltage peak occur?

    A. Zero times
    B. One time
    C. Two times
    D. Four times

_____ 18. The standard frequency for alternating current in the United States is _____.

    A. 60 Hz
    B. 120 Hz
    C. 1800 Hz
    D. 3600 Hz

Name _____

12.3 Electrical Materials

_____ 19. Conductors are substances that allow unrestricted current flow because _____.

    A. their atoms have free electrons
    B. they are all magnetized
    C. they are naturally well lubricated
    D. All of the above.

_____ 20. Insulators are substances that do not allow current flow because _____.

    A. their atoms have no free electrons
    B. they are not magnetized
    C. they are not naturally well lubricated
    D. All of the above.

_____ 21. A semiconductor may be triggered to function as a conductor by _____.

    A. an electrical signal
    B. light intensity
    C. temperature
    D. All of the above.

_____ 22. *True or False?* Copper and iron are both electrical conductors, but copper has higher resistance than iron.

12.4 Circuit Fundamentals

Match the following electrical symbols with the device or component name that the symbol represents.

_____ 23. Capacitor

_____ 24. Flow-activated switch

_____ 25. Fusible link

_____ 26. Ground connection

_____ 27. Liquid level switch

_____ 28. Motor

_____ 29. Pressure-activated switch

_____ 30. Temperature-activated switch

_____ 31. Thermistor

_____ 32. Transformer

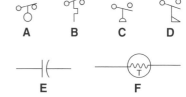

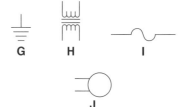

Goodheart-Willcox Publisher

_____ 33. *True or False?* In a series circuit, the current through each load is added together to calculate the total series current.

_____ 34. *True or False?* In a parallel circuit, the current through each load is identical, regardless of each load's resistance.

_____ 35. What is the voltage drop across a single electrical load that has a resistance of 2 Ω and a current of 5 A flowing through it?

A. 2.5 V
B. 3 V
C. 7 V
D. 10 V

_____ 36. The electrical loads in _____ circuits each have the same voltage drop, which equals the total applied voltage of the circuit.

A. series
B. parallel
C. series-parallel
D. All of the above.

12.5 Magnetism

_____ 37. Regarding magnetism, like poles _____.

A. attract each other
B. cause one of the poles to reverse orientation
C. ignore each other
D. repel each other

_____ 38. When a material becomes a magnet because it is placed in a magnetic field, this phenomenon is called _____.

A. electromagnetism
B. induced magnetism
C. parallel magnetism
D. secondary magnetism

_____ 39. When a conductor wound around a piece of soft iron conducts current and turns the soft iron into a magnet, this phenomenon is called _____.

A. current magnetism
B. electromagnetism
C. secondary magnetism
D. voltage magnetism

_____ 40. *True or False?* If more coil turns are added to the winding of an electromagnet, the strength of the electromagnet's magnetic field will increase.

Name _____

12.6 Electrical Generators

_____ 41. Electrical generators produce voltage by moving a conductor ____.
   A. across a magnetic field
   B. along a rugged surface
   C. in a reciprocating motion in a cylinder
   D. through a narrow semiconductor orifice

_____ 42. An ac generator uses a(n) ____ to transmit current from the wire loop.
   A. brush wheel
   B. commutator
   C. iron core
   D. slip ring

_____ 43. A dc generator uses a(n) ____ to transmit current from the wire loop.
   A. brush wheel
   B. commutator
   C. iron core
   D. slip ring

12.7 Transformer Basics

_____ 44. A transformer can only operate using ____.
   A. alternating current
   B. cores made of a dielectric material
   C. direct current
   D. oil-filled capacitors

_____ 45. In transformer operation, a magnetic field is initially generated by current in the ____.
   A. brushes
   B. core
   C. primary coil
   D. secondary coil

Critical Thinking

_____ 46. What is the resistance of a circuit if it has a voltage of 50 V and a current of 8 A? Show your calculations.

_____ 47. What is the voltage of a circuit with a current of 12 A and a resistance of 6 Ω?
Show your calculations.

_____ 48. What is the current in a circuit if the voltage is 120 V and the resistance is 15 Ω.
Show your calculations.

_____ 49. If a circuit has a current of 6 A and a voltage of 100 V, by how much must the resistance
increase for the current to reduce to 5 A? Show your calculations.

Name _____

50. List at least five materials that are commonly used as electrical insulators. Explain how these materials are used in electrical components.

_____

_____

_____

_____

_____

51. Why are electronic devices that are made up of semiconductors referred to as solid-state devices?

_____

_____

_____

5

_____ 52. If a series circuit has three loads, each with a resistance of 6 Ω, and the power source provides 120 V, what is the current flowing through each of the loads? Show your calculations.

53. List four factors that affect the strength of an electromagnet.

_____

_____

_____

_____

_____ 54. If the primary voltage of a transformer is 240 V and the primary coil has 300 turns of wire, how many turns of wire would be required in the secondary coil to create a secondary voltage of 40 V? Show your calculations.

_____ 55. What is the output voltage of a transformer that has a primary coil with 600 turns, a secondary coil with 450 turns, and a primary voltage of 120 V? Show your calculations.

Electrical Power

5

| Carefully study the chapter and then answer the following questions. |

13.1 Electrical Power

_____ 1. When current flows through an electrical component due to a potential difference, electrical _____ is produced.

    A. capacitance
    B. ground
    C. power
    D. resistance

_____ 2. How many watts does an electrical load use if a current of 1 A flows through it when the load is connected to a 1 V power source?

    A. 0 W
    B. 1 W
    C. 2 W
    D. 10 W

_____ 3. The effective voltage of an alternating current is calculated using the formula _____.

    A. $I \times E$
    B. $I^2 \times 0.707$
    C. $I^2 \times R$
    D. $V_{max} \times 0.707$

_____ 4. _True or False?_ Unintentional electrical resistance may result from improperly sized conductors.

_____ 5. _True or False?_ Resistance causes a phase shift in which current lags behind voltage.

_____ 6. The electrical property that opposes a change in current is called _____.

    A. capacitance
    B. inductance
    C. resistance
    D. wattage reactance

_____ 7. In an electrical circuit, voltage may be caused to lag behind current due to _____.
   A. capacitive reactance
   B. inductive reactance
   C. resistive reactance
   D. wattage reactance

_____ 8. In an electrical circuit, current may be caused to lag behind voltage due to _____.
   A. capacitive reactance
   B. inductive reactance
   C. resistive reactance
   D. wattage reactance

_____ 9. _True or False?_ If an electrical circuit is purely resistive, its apparent power cannot equal its true power.

_____ 10. _True or False?_ In an electrical circuit, inductive components can cause voltage and current to be out of phase.

_____ 11. The ratio of true power to apparent power is the _____ of a circuit.
   A. grounding factor
   B. overload factor
   C. median square factor
   D. power factor

_____ 12. The actual power used by a circuit is the _____ power of the circuit.
   A. apparent
   B. capacitive
   C. inductive
   D. true

_____ 13. If the true power and apparent power are _not_ equal, the power factor of a circuit is _____.
   A. greater than 100%
   B. less than 100%
   C. incorrect
   D. None of the above.

13.2 Power Circuits

_____ 14. In a three-phase voltage cycle, three separate voltage signals are delayed so they peak at different _____.
   A. amperages
   B. frequencies
   C. times
   D. values

_____ 15. Local electrical codes often use standards set by the _____.
   A. IFGC
   B. IMC
   C. LOTO
   D. NEC

_____ 16. _True or False?_ A Class 2 circuit is defined as a circuit supplied by a power source that has an output no greater than 30 V and 1000 VΛ.

Name _____

_____ 17. *True or False?* Class 2 circuits do *not* include remote-control circuits with a relay, thermostat, or any other device that controls another circuit.

_____ 18. As a conductor's wire diameter increases, its AWG number _____.

  A. decreases
  B. increases
  C. remains constant
  D. None of the above.

_____ 19. As wire diameter size increases, its electrical resistance _____.

  A. decreases
  B. increases
  C. remains constant
  D. None of the above.

_____ 20. The numbers for aught size wires increase as the wire size _____.

  A. decreases
  B. increases
  C. remains constant
  D. None of the above.

_____ 21. *True or False?* There is no difference between a 4 AWG wire and a 4/0 AWG wire.

Match each letter under the electrical wiring terminals shown with its proper name.

_____ 22. Flanged spade

_____ 23. Flag

_____ 24. Hook

_____ 25. Ring

_____ 26. Spade

A   B   C   D   E

Goodheart-Willcox Publisher

_____ 27. An overcurrent protection device that detects current imbalances between ungrounded and grounded conductors is a _____.

  A. circuit breaker
  B. fuse
  C. ground fault circuit interrupter
  D. thermistor

_____ 28. An overcurrent protection device that regulates current by changing its resistance based on heat is a _____.

  A. circuit breaker
  B. fuse
  C. ground fault circuit interrupter
  D. thermistor

_____ 29. An overcurrent protection device that uses a solenoid to open the circuit when current rises to a preset limit is a _____.

   A. circuit breaker
   B. fuse
   C. ground fault circuit interrupter
   D. thermistor

_____ 30. An overcurrent protection device that is designed to heat up and melt to stop high current is a _____.

   A. circuit breaker
   B. fuse
   C. ground fault circuit interrupter
   D. thermistor

Match each electrical term with the phrase that best describes it.

_____ 31. The neutral wires connected to a transformer.

_____ 32. The creation of a continuous electrical connection of all metal parts in an electrical system.

_____ 33. The soil to which an electrical system connects.

_____ 34. The hot wires connected to the phase lines of a transformer.

_____ 35. The connection of an electrical system to the earth or the equipment that connects the system to the earth.

A. Bonding

B. Ground

C. Grounded conductors

D. Grounding

E. Ungrounded conductors

13.3 Electrical Problems

_____ 36. A(n) _____ occurs when a break in the electrical path stops a circuit's current.

   A. ground fault
   B. open circuit
   C. overload
   D. short circuit

_____ 37. A(n) _____ occurs when current is routed around an electrical load instead of through it.

   A. ground fault
   B. open circuit
   C. short circuit
   D. unintentional voltage drop

_____ 38. A(n) _____ is a condition in which too much current flows through a circuit.

   A. ground fault
   B. open circuit
   C. overload
   D. unintentional voltage drop

Name _____

_____ 39. A(n) _____ is a condition in which a current-carrying conductor touches a grounded object.
   A. ground fault
   B. open circuit
   C. overload
   D. unintentional voltage drop

_____ 40. A(n) _____ is a condition in which a circuit's voltage is reduced without intent or outside of the normal design.
   A. ground fault
   B. overload
   C. short circuit
   D. unintentional voltage drop

5

Critical Thinking

_____ 41. What is the effective voltage (rms voltage) of an ac power source with a maximum voltage of 150 V? Show your calculations.

_____ 42. If the maximum current of an ac power source peaks at 6 A, what is the effective current? Show your calculations.

_____ 43. Calculate the rms current of an ac power source with a maximum current that peaks at 8 A. Show your calculations.

_____ 44. Using Watt's law, what is the electrical power of a circuit that has a 6 A current and a 24 Ω resistance? Show your calculations.

_____ 45. Using Watt's law, what is the electrical power of a circuit that has a 20 A current and a 24 Ω resistance? Show your calculations.

_____ 46. Calculate the apparent power of a circuit with a voltmeter reading of 75 V and an ammeter reading of 10 A. Show your calculations.

Name _____

_____ 47. What is the power factor of an ac circuit with an apparent power of 1200 VA and true power of 900 W? Show your calculations.

48. How might the power factor of inductive circuits be improved?

_____

_____

_____

_____

49. List four variables that affect the compatibility of an electrical system with the power provided by the electric utility company. Include common examples of each used in residential and commercial applications.

_____

_____

_____

_____

50. Explain why it is important for a power supply to match the electrical specifications of an electrical device.

_____

_____

_____

_____

51. List at least three causes of overloads and provide brief explanations of why these might occur.

_____

_____

_____

_____

_____

_____ 52. Calculate the voltage drop of a 20′ conductor carrying 25 A with a resistance of 1.5 Ω. Use the following formula: voltage drop = $(2 \times L \times R \times I) \div 1000$. Show your calculations.

53.   List at least three different things that can cause an open circuit.

_____

_____

_____

CHAPTER 14

Name _____

Date _____ Class _____

Basic Electronics

5

> *Carefully study the chapter and then answer the following questions.*

14.1 Semiconductor Basics

> *Match the following terms with the phrase that best describes them.*

_____ 1. Substance that can be made to conduct electricity under certain conditions

_____ 2. The addition of impurities to a pure semiconductor to alter its conductivity

_____ 3. Name of a theory explaining how electrons and positively charged spaces move through a semiconductor

_____ 4. Negatively charged terminal of some semiconductors

_____ 5. Positively charged side of some semiconductors

A. Anode

B. Cathode

C. Doping

D. Hole flow

E. Semiconductor

_____ 6. *True or False?* Electrons flow through a diode from the anode to the cathode.

_____ 7. *True or False?* Diodes may also be called rectifiers because they may be used to convert ac to dc in a circuit.

_____ 8. Doping produces _____ material when the impurity causes a surplus of electrons.
A. D-type
B. N-type
C. P-type
D. Z-type

_____ 9. Doping produces _____ material when the impurity causes a shortage of electrons.
A. D-type
B. N-type
C. P-type
D. Z-type

_____ 10. A diode is forward biased when the negative terminal from a dc power supply is connected to the diode's ____.

   A. anode
   B. cathode
   C. base
   D. gate

_____ 11. What happens when a diode is reverse biased?

   A. Current will flow through the circuit.
   B. Current will not flow through the circuit.
   C. The diode burns out.
   D. An overload protection device functions.

14.2 Control Circuits and Electronic Devices

_____ 12. A circuit in which electrical or electronic devices are used to regulate current flow is called a ____ circuit.

   A. control
   B. sensor
   C. signal
   D. transducer

_____ 13. *True or False?* A major difference between electronic circuits and electrical circuits is that an electrical circuit uses semiconductors and solid-state devices control current.

Match the letter of the schematic symbol to the electronic devices listed.

_____ 14. Diode

_____ 15. Diac

_____ 16. SCR

_____ 17. Transistor

_____ 18. Triac

Goodheart-Willcox Publisher

_____ 19. A device that detects and responds to some kind of stimulus, such as changes in temperature or pressure, is a(n) ____.

   A. inverter
   B. sensor
   C. transducer
   D. transistor

Name _____

_____ 20. A device that converts an input signal from one form of energy to an output signal of another form of energy is a(n) _____.

A. inverter
B. sensor
C. transducer
D. transistor

_____ 21. A two-terminal solid-state device that allows current to flow in both directions once breakover voltage is reached and so long as holding current remains high enough is a(n) _____.

A. diac
B. inverter
C. SCR
D. triac

_____ 22. A three-terminal solid-state device that allows current to flow in only one direction once breakover voltage is reached or voltage is applied to its gate is a(n) _____.

A. inverter
B. SCR
C. transistor
D. triac

_____ 23. A three-terminal solid-state device that allows current to flow in both directions once voltage is applied to its gate is a(n) _____.

A. diac
B. inverter
C. SCR
D. triac

_____ 24. A three-terminal solid-state device that functions as a switch or an electrical signal amplifier is a(n) _____.

A. inverter
B. SCR
C. transistor
D. triac

_____ 25. A solid-state circuit that converts dc to ac is a(n) _____.

A. diac
B. inverter
C. transducer
D. transistor

_____ 26. A positive temperature coefficient thermistor decreases its resistance as _____.

A. temperature drops
B. temperature rises
C. Both A and B.
D. Neither A nor B.

_____ 27. A negative temperature coefficient thermistor decreases its resistance as _____.

A. temperature drops
B. temperature rises
C. Both A and B.
D. Neither A nor B.

_____ 28. A semiconductor device that produces light when a voltage is applied to it is a ____ device.

    A. photoconductor
    B. photoemissive
    C. photographic
    D. photovoltaic

_____ 29. A semiconductor device that increases or decreases its resistance to electric current based on its exposure to electromagnetic radiation is a ____ device.

    A. photoconductor
    B. photoemissive
    C. photographic
    D. photovoltaic

_____ 30. A semiconductor device that produces energy when it is exposed to light is a ____ device.

    A. photoconductor
    B. photoemissive
    C. photographic
    D. photovoltaic

14.3 Circuit Boards and Microprocessors

_____ 31. What completes the circuit between the electronic devices found on a printed circuit board?

    A. Conductive metal strips
    B. Directly connecting their terminals
    C. Radio signals
    D. Wiring

_____ 32. A single device capable of accepting information, storing it, and reacting in some preset way describes a(n) ____.

    A. inverter
    B. microprocessor
    C. NTC thermistor
    D. PTC thermistor

14.4 Switches and Contacts

_____ 33. The physical parts of switches and relays that complete an electric circuit are called ____.

    A. contacts
    B. junctions
    C. poles
    D. throws

Name _____

Match the letter of the switch arrangements with their descriptions.

_____ 34. Single-pole, single-throw

_____ 35. Single-pole, double-throw

_____ 36. Double-pole, single-throw

_____ 37. Double-pole, double-throw

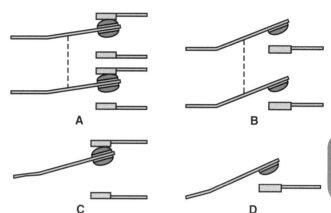

Micro Switch, Div. of Honeywell, Inc.

14.5 Relays

_____ 38. A relay is an electrical switching device that operates under the control of _____.

    A. a beam of light

    B. an outside electrical signal

    C. pressure

    D. temperature

_____ 39. With power applied to a relay's armature (common) terminal and also across its coil terminals, an electrical load can be powered by the _____ terminal.

    A. NC

    B. NO

    C. Both terminals.

    D. Neither terminals.

_____ 40. With power applied only to a relay's armature (common) terminal, an electrical load can be powered by the _____ terminal.

    A. NC

    B. NO

    C. Both terminals.

    D. Neither terminals.

Match the following components of a relay with the corresponding letters in the image below.

_____ 41. Armature

_____ 42. Armature terminal

_____ 43. Coil

_____ 44. Coil terminals

_____ 45. NC contact

_____ 46. Normally closed (NC) terminal

_____ 47. Normally open (NO) terminal

_____ 48. Spring

Goodheart-Willcox Publisher

14.6 Solenoids

_____ 49. Solenoids are used to turn electrical energy into _____.

A. chemical energy
B. mechanical motion
C. thermal energy
D. visible light

_____ 50. A solenoid that uses gravity to return its plunger back into position when de-energized must be mounted _____.

A. sideways
B. upright
C. upside down
D. with its coil removed

14.7 Thermocouples

_____ 51. The concept underlying the conversion of thermal energy into electrical power is the _____ effect.

A. Celsius
B. Fahrenheit
C. Peltier
D. Seebeck

Name _____

_____ 52. The concept that explains how electric current is used to transfer heat from one place to another is the _____ effect.
A. Celsius
B. Fahrenheit
C. Peltier
D. Seebeck

Critical Thinking

53. In your own words, explain what electronic devices are.

_____

_____

_____

_____

54. In your own words, describe what a printed circuit board is. List a few places where printed circuit boards can be found in different HVACR systems.

_____

_____

_____

_____

55. How do printed circuit boards simplify the servicing of HVACR systems?

_____

_____

_____

_____

56. In your own words, explain what a thermocouple is. How are they used in HVACR systems?

_____

_____

_____

_____

Notes

CHAPTER 15

Electric Motors

Name _____

Date _____ **Class** _____

6

Carefully study the chapter and then answer the following questions.

15.1 The Elementary Electric Motor

_____ 1. An electric motor operates based on the principles of electricity and _____.

A. hydraulics
B. magnetism
C. pneumatics
D. radiation

_____ 2. The moving part inside an electric motor is the _____.

A. end bell
B. rotor
C. stator
D. terminal box

_____ 3. The stationary part inside an electric motor is the _____.

A. end bell
B. rotor
C. stator
D. terminal box

_____ 4. A(n) _____ is an electromagnet whose polarity changes as the flow of current alternates in the field windings.

A. end bell
B. field pole
C. field windings
D. terminal box

_____ 5. Electrical connections are made in the _____ to control and power the motor.

A. field pole
B. field windings
C. rotor
D. terminal box

_____   6. The wires wrapped around the field poles of the stator are called _____.
   A. end bell
   B. field pole
   C. field windings
   D. terminal box

The following statements refer to the basic operation of an electric motor. Determine which of the terms listed on the right best completes each statement. Not all letters will be used.

_____   7. When current from a power source flows through the field windings, it creates a(n) _____.

_____   8. The stator's poles switch _____ with the applied alternating current.

_____   9. As the polarities of the stator's poles change due to the flow of _____, the polarity of the rotor also changes.

_____   10. Current induced in the rotor creates an electromagnetic field that interacts with the electromagnetic field created by the _____.

_____   11. The attraction of unlike poles and _____ of like poles causes the rotor to rotate.

_____   12. The magnetic field created in the rotor has a polarity _____ of the polarity of the magnetic field in the stator windings.

_____   13. The interaction between the magnetic fields of the stator and rotor causes the rotor to _____.

A. alternating current

B. electromagnetic field

C. induction

D. opposite

E. polarity

F. repulsion

G. rotate

H. stator

I. synchronous

_____   14. The oppositely polarized voltage induced by the magnetic field in the rotor is called _____ force.
   A. back magnetomotive
   B. counter electromotive
   C. reciprocative
   D. retromotive

_____   15. *True or False?* The speed of dc motors is dependent primarily on the current's frequency and the number of poles in the motor.

_____   16. *True or False?* Electric motors are rated at their operating speed under a full load.

_____   17. Mechanical energy produced by the motor shaft divided by the power input to the motor is referred to as motor _____.
   A. cemf
   B. efficiency
   C. duty
   D. slip

Name _____

_____ 18. The difference between synchronous speed and rated full-load speed is a motor's ____.

    A. cemf

    B. efficiency

    C. duty

    D. slip

15.2 AC Induction Motors

_____ 19. Induction motors are ac motors that operate by using the magnetic field generated in the stator to ____.

    A. change the frequency of the electrical signal

    B. change the number of motor poles

    C. induce current in the rotor

    D. All of the above.

_____ 20. _True or False?_ Run windings are used for motor starting and additional torque.

_____ 21. The start winding has a higher resistance than the run winding because it is made of ____.

    A. copper instead of iron

    B. larger gage wire

    C. permanent magnets

    D. smaller gage wire

_____ 22. The start winding has a higher inductance than the run winding because it has ____.

    A. fewer coil turns

    B. larger gage wire

    C. more coil turns

    D. more magnets

_____ 23. _True or False?_ When a split-phase motor is starting, current is going through both the start and run windings.

_____ 24. What happens when a single-phase motor's start winding is left in the circuit when the motor reaches 60% to 75% of rated full-load speed?

    A. The motor continues running with no problems.

    B. Motor speed drops to 50% rated full-load speed.

    C. The run winding drops out of the circuit.

    D. The start winding may overheat.

_____ 25. Capacitors cause capacitive ____ in a circuit, which is the opposition to the flow of current that causes voltage to lag behind current in an ac cycle.

    A. flow

    B. induction

    C. lag

    D. reactance

_____ 26. _True or False?_ A start capacitor is used in a stator's run winding circuit.

_____ 27. A larger phase displacement leads to a higher starting torque, which means motors that use capacitors can start _____.

  A. a little earlier than normal
  B. slow and gradually ramp up to speed
  C. under heavier loads
  D. with no slip

_____ 28. Split-phase motors are very popular in compressors in HVACR systems that use a(n) _____ as the metering device.

  A. capillary tube
  B. EEV
  C. low-side float
  D. TXV

_____ 29. The contacts of a _____ change position when force applied to weights overcome spring pressure.

  A. centrifugal switch
  B. ECM
  C. start relay
  D. VFD

Identify the parts of the CSCR motor circuit shown in the following illustration.

_____ 30. Centrifugal switch

_____ 31. Run winding

_____ 32. Run capacitor

_____ 33. Start capacitor

_____ 34. Start winding

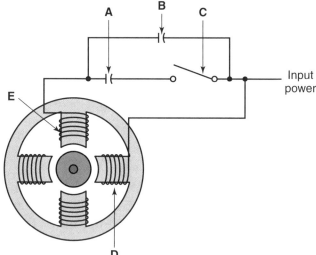

Goodheart-Willcox Publisher

Name _____

Determine which of the terms listed on the right best completes each statement. Not all letters will be used.

_____ 35. A device used to reduce a compressor's load on start-up is called a(n) ____.

_____ 36. A(n) ____ is a single-phase induction motor that has a start capacitor wired in series with the start winding.

_____ 37. A(n) ____ is a single-phase induction motor that has a start capacitor and a run capacitor wired in series with the start winding.

_____ 38. A(n) ____ is a single-phase induction motor that uses a single run capacitor in series with the start winding for the motor's entire operation.

_____ 39. A(n) ____ is a low-torque, single-phase motor that has field poles that are split with part of each wrapped in a copper band to produce starting torque.

_____ 40. A(n) ____ is an induction motor with three sets of stator windings operating on three separate power signals.

_____ 41. A(n) ____ has its stator windings arranged in pairs so that it can be used at two different voltages.

A. capacitor-start, capacitor-run (CSCR) motor

B. capacitor-start, induction-run (CSIR) motor

C. common

D. dual-voltage motor

E. permanent split capacitor (PSC) motor

F. shaded-pole motor

G. three-phase motor

H. unloader

I. VFD

J. windings

6

_____ 42. *True or False?* VFDs are a great option to use when controlling the speed of dc motors.

_____ 43. A VFD controls the speed of motors by changing the ____.
   A. frequency of the electrical signal
   B. level of applied voltage
   C. motor's ambient conditions
   D. number of poles in the motor

15.3 Electronically Commutated Motors (ECMs)

_____ 44. ECMs have rotors made of ____ instead of coils of wire or metal bars.
   A. crystals
   B. hollow wire cages
   C. permanent magnets
   D. quartz

_____ 45. *True or False?* A major maintenance issue with ECMs is that their brushes wear out frequently and need replacing.

_____ 46. *True or False?* ECMs usually start quietly but maintain an overwhelming roar during normal operation.

15.4 Standard Motor Data

_____ 47. The current level at which a motor's full-load torque and horsepower are reached is called the ____ amperage.

A. continuous duty
B. inherent motor
C. locked rotor
D. full-load

_____ 48. The amount of mechanical work that a motor can perform based on both motor speed and torque is the rated ____.

A. horsepower
B. power factor
C. service factor
D. voltage

_____ 49. The letter and number designation that indicates the physical dimensions of a motor is its ____ size.

A. frame
B. inherent
C. phase
D. service

_____ 50. NEC terms "continuous duty" and "intermittent duty" primarily deal with a motor's ____ rating.

A. horsepower
B. phase
C. service
D. time

_____ 51. Two prominent forms of inherent motor protection are ____.

A. magnetically protected and impedance protected
B. magnetically protected and thermally protected
C. impedance protected and thermally protected
D. None of the above.

15.5 Motor Applications in HVACR Systems

_____ 52. Motors that start under load need ____.

A. higher starting torque
B. less current
C. lower starting torque
D. smaller conductors

_____ 53. *True or False?* One method of changing the speed of a belt-driven compressor is by replacing the pulley with one of a different size.

Name _____

_____ 54. A hermetic compressor's motor _____.

    A. requires leakproof electrical connections
    B. requires special cooling provisions
    C. Both A and B.
    D. Neither A nor B.

_____ 55. *True or False?* Motor-driven fans used to circulate air around a condenser are often wired to run when the compressor runs.

Critical Thinking

_____ 56. What is the synchronous speed of a two-pole motor that operates at a frequency of 50 Hz? Use the following formula: $120 \times (\text{frequency} \div \text{poles})$. Show your calculations.

_____ 57. What is the synchronous speed of a four-pole motor that operates at a frequency of 50 Hz? Show your calculations.

58. In your own words, explain what "phase splitting" is. What is it used to do?

_____

_____

_____

_____

_____

_____

59. In your own words, explain what should happen when a single-phase induction motor reaches 60% to 75% of its rated full-load speed. What happens after that?

_____

_____

_____

60. List and explain four advantages of ECMs.

_____

_____

_____

_____

61. Why is it important to select a motor with the correct locked rotor code letter?

_____

_____

_____

62. If the shaft of a belt-driven compressor is turning at one speed, how is it that the shaft of its motor is turning at a different speed? Explain this in your own words.

_____

_____

_____

Electrical Control Systems

Carefully study the chapter and then answer the following questions.

16.1 Circuit Diagrams

_____ 1. To understand the order of device operation in a system, use a _____ diagram.
   A. ladder
   B. pictorial
   C. P/T
   D. Venn

_____ 2. To see how electrical devices are connected inside a unit, use a _____ diagram.
   A. ladder
   B. pictorial
   C. P/T
   D. Venn

_____ 3. To locate the approximate physical location of a device quickly, use a _____ diagram.
   A. ladder
   B. pictorial
   C. P/T
   D. Venn

_____ 4. To know the color of wiring running to and from devices in a unit, use a _____ diagram.
   A. ladder
   B. pictorial
   C. P/T
   D. Venn

_____ 5. When troubleshooting a system that is not operating properly, use a _____ diagram.
   A. ladder
   B. pictorial
   C. P/T
   D. Venn

_____ 6. Each horizontal line on a ladder diagram represents a(n) ____.

    A. control component
    B. individual circuit with a load
    C. power wire
    D. switch

16.2 Control System Fundamentals

_____ 7. The condition value at which a device ceases to operate is its ____.

    A. cut-in
    B. cut-out
    C. differential
    D. range

_____ 8. Pressure sensors use a(n) ____ to change their pressure signal into a mechanical signal.

    A. crystal
    B. diaphragm or bellows
    C. electromagnet
    D. thermistor

_____ 9. The number of units between when a device begins operation and ceases operation is its ____.

    A. cut-in
    B. cut-out
    C. differential
    D. range

_____ 10. A common mode of communication used in control systems is ____.

    A. hydraulic
    B. pneumatic
    C. electrical
    D. All of the above.

_____ 11. *True or False?* Signals sent by a controller may be used to change a system's operating parameters.

_____ 12. The condition value at which a device begins to operate is its ____.

    A. cut-in
    B. cut-out
    C. differential
    D. range

_____ 13. The set of numbers between and including cut-in and cut-out values is a device's ____.

    A. cut-in
    B. cut-out
    C. differential
    D. range

Name _____

_____ 14. If a commercial cooler's temperature is rising too high but should *not* be adjusted lower (to protect sensitive products), a technician should adjust the _____ control.

    A. differential combination control
    B. differential cut-in control
    C. differential cut-out control
    D. range adjustment

_____ 15. If an HVACR system's average temperature was acceptable but its high was too high and low was too low, which one control should be adjusted to fix this?

    A. Differential combination control
    B. Differential cut-in control
    C. Differential cut-out control
    D. Range adjustment

_____ 16. If an HVACR system was overall too warm and all values needed to be reduced at once in equal amount, which one control should be adjusted?

    A. Differential combination control
    B. Differential cut-in control
    C. Differential cut-out control
    D. Range adjustment

16.3 Motor Controls

Match the following terms with the phrase that best describes them.

_____ 17. A sensing device in an electronic temperature sensor

_____ 18. A characteristic in which resistance decreases as the temperature increases

_____ 19. Used to shut down the compressor in the event of refrigerant loss

_____ 20. Similar to a contactor, but it has built-in overload protection

_____ 21. A type of relay that operates using only electronic devices

_____ 22. A type of relay that uses a thermistor to protect motor circuits

_____ 23. Used to detect changes in a thermistor's resistance

A. Low-side pressure safety control

B. Motor starter

C. Negative temperature coefficient (NTC)

D. Positive temperature coefficient (PTC)

E. Solid-state

F. Thermistor

G. Wheatstone bridge

_____ 24. A sensing bulb is primarily used by _____ motor control

    A. bimetal
    B. electronic
    C. pressure
    D. thermostatic

_____ 25. In order to produce a snap action, a bimetal strip is formed into a(n) _____.

    A. coil
    B. disc
    C. helix
    D. S-shape

_____ 26. For an oil pressure motor control to allow an HVACR system to continue operating, the oil pressure sensed must be _____.

    A. above low-side pressure
    B. below low-side pressure
    C. the same as low-side pressure
    D. in vacuum

_____ 27. Which of the following is a type of temperature sensor used with a thermostatic motor control?

    A. Bimetal device
    B. Electronic sensor
    C. Sensing bulb
    D. All of the above.

_____ 28. A high-pressure motor control is connected to the _____.

    A. accumulator
    B. compressor outlet
    C. liquid receiver
    D. suction line

_____ 29. Relays can be used to _____.

    A. provide power directly to motors
    B. add start windings to a motor circuit
    C. add start capacitors to a motor circuit
    D. All of the above.

_____ 30. *True or False?* Upon reaching rated full-load speed, a start capacitor is often added to a single-phase motor circuit.

_____ 31. Potential relays operate based on _____.

    A. cemf
    B. changing temperature
    C. high inrush of current
    D. sales potential

_____ 32. The required voltage in the start winding necessary to energize a potential relay coil is called _____ voltage.

    A. breakover
    B. go-time
    C. pickup
    D. holding

_____ 33. *True or False?* Most starting relays can be and often are rebuilt in an HVAC shop for later reuse.

Name _____

_____ 34. Current relays operate based on ____.

    A. cemf

    B. changing temperature

    C. high inrush of current

    D. sales potential

_____ 35. A ____ relay uses a weight or a spring to hold the start winding contact points open when the system is idle.

    A. current

    B. potential

    C. PTC

    D. solid-state

Match each part listed with its corresponding letter in the photograph.

_____ 36. Grounding wire screw

_____ 37. Cut-in adjustment screw

_____ 38. Capillary tube

_____ 39. Cut-in indicator

_____ 40. Connects to the low side of the refrigerant circuit

_____ 41. Differential indicator

_____ 42. Differential adjustment screw

_____ 43. Pressure element

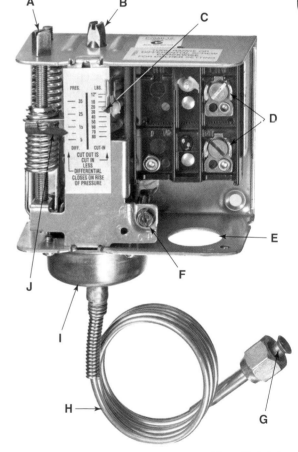

Johnson Controls, Inc.

16.4 Motor Protection Devices

Match each of the following motor protection devices with the phrase that best describes them.

_____ 44. Opens a circuit when the current exceeds its limit and closes when it cools down

_____ 45. Increases resistance to prevent current from conducting when temperature rises beyond a safe level

_____ 46. A piece of metal that melts to open the circuit when current rises too high

_____ 47. May only open a circuit if the current is too high and *not* if the motor is overheating

_____ 48. A dual-element device that will *not* blow unless an overload condition exists for a certain period of time

_____ 49. A device that is fast-acting but also has a time-delay feature

_____ 50. An automatic switch that opens a circuit if the current draw exceeds a predetermined level and must be manually reset

_____ 51. A device that blows immediately after its maximum rating is exceeded

A. Automatic reset

B. Circuit breaker

C. Current-limiting fuse

D. External bimetal protection device

E. Fast-acting fuse

F. Fuse

G. Multipurpose fuse

H. PTC thermistor

I. Time-delay fuse

_____ 52. The internal protection device in a three-phase motor will open _____.
   A. all three phases
   B. the common
   C. phases 1 and 2
   D. phases 2 and 3

_____ 53. Most motors normally operate at _____.
   A. 100°F (38°C)
   B. 125°F (52°C)
   C. 200°F (93°C)
   D. 250°F (121°C)

16.5 Direct Digital Control (DDC)

_____ 54. Most wiring for DDC is intended to carry _____.
   A. 0–10 mVdc
   B. 0–10 Vdc
   C. 30–50 Vdc
   D. 110–120 Vdc

_____ 55. *True or False?* A closed-loop control system is closed because it does detect or use feedback from the conditioned space.

_____ 56. *True or False?* A variable-position damper is an example of an on-off control.

> *Match the following terms with the phrase that best describes them.*

_____ 57. The difference between the present condition and the desired condition

_____ 58. The opposition to the flow of alternating current

_____ 59. The desired condition in an area

_____ 60. Information that is detected by a sensor in a conditioned area and sent to the controller

_____ 61. A device that operates based on signals from another device

_____ 62. A device that converts an input signal of one form of energy into an output signal of another form of energy

_____ 63. The present condition as measured by sensors

_____ 64. A controlled device that changes an input signal into mechanical motion

_____ 65. A circuit that responds to changes in signals from sensors and issues signals in response

_____ 66. A device that detects a specific variable in a conditioned space

A. Actuator

B. Controlled device

C. Controller

D. Control point

E. Feedback

F. Impedance

G. Offset

H. Sensor

I. Set point

J. Transducer

6

Critical Thinking

67. Explain what an open-loop control system is. How is its operation changed?

_____

_____

_____

_____

68. Explain what a closed-loop control system is.

_____

_____

_____

_____

69. Explain why an above-atmospheric-pressure element that has lost its charge will not be able to activate its switch.

_____

_____

70. Why must contact points operated by a sensing bulb's bellows or diaphragm snap open and closed rapidly?

_____

_____

_____

71. List three advantages electronic temperature sensors have over other temperature sensors.

_____

_____

_____

72. List three conditions that cause high head pressure and may result in motor shutdown.

_____

_____

_____

_____

_____

73. List five factors that may cause an oil pressure motor control to trip open.

_____

_____

_____

_____

_____

74. Calculate the current levels that would be applied to a normally open damper to turn it into the following positions. Assume that the DDC control signals are along a 4 mA to 20 mA scale.

Fully open: _____

25% closed: _____

50% closed: _____

75% closed: _____

100% closed: _____

75. What is a lockout relay? How does it protect a circuit? How is normal operation restored?

_____

_____

_____

_____

CHAPTER 17

Name _____

Date _____ Class _____

Servicing Electric Motors and Controls

> *Carefully study the chapter and then answer the following questions.*

17.1 Electrical Test Equipment

_____ 1. A multimeter can be used to measure _____.

    A. current

    B. resistance

    C. voltage

    D. All of the above.

_____ 2. Meters designed to measure potential difference between two points in a circuit are _____.

    A. megohmmeters

    B. ohmmeters

    C. voltmeters

    D. wattmeters

_____ 3. *True or False?* To ensure that an ohmmeter reading will be across the intended device and not the rest of the circuit, a technician should disconnect at least one leg of the device.

_____ 4. *True or False?* Before using an ohmmeter, use a meter on a circuit to ensure that no voltage is present.

_____ 5. If the current is too low for a(n) _____ to measure and the wire is long enough, the wire can be wrapped multiple times to get a higher reading.

    A. clamp-on ammeter

    B. meghommeter

    C. ohmmeter

    D. voltmeter

_____ 6. In a dc circuit, an in-line ammeter's _____ lead is placed on the side of the load closest to the negative terminal of the circuit's power supply.

    A. black

    B. green

    C. red

    D. white

_____ 7. A(n) _____ does *not* require the disconnection of wires or the attachment of leads to obtain a reading.

    A. clamp-on ammeter
    B. meghommeter
    C. ohmmeter
    D. voltmeter

> *Match the functions and settings on the digital multimeter with the terms listed below.*

_____ 8. Diode check

_____ 9. Common terminal (black lead)

_____ 10. Low-current terminal—below 1 A (red lead)

_____ 11. Volts dc

_____ 12. Microamps ac

_____ 13. Capacitance

_____ 14. Voltage, resistance, capacitance, and frequency terminal (red lead)

_____ 15. High-current terminal—above 1 A (red lead)

_____ 16. Resistance

_____ 17. Frequency check

_____ 18. Volts ac

_____ 19. Amps ac

_____ 20. Millivolts dc

_____ 21. Milliamps ac

_____ 22. Continuity check

Sealed Unit Parts Co., Inc.

_____ 23. *True or False?* The black lead is always connected to one of the positive (+) terminals on a multimeter.

_____ 24. *True or False?* A diode must be replaced if a meter registers a voltage for both lead configurations.

Name _____

_____ 25. A power factor meter _____.

    A. determines the apparent power of a circuit

    B. measures the circuit's continuity

    C. measures extended flow

    D. All of the above.

_____ 26. *True or False?* Burnt or heat-damaged motor winding insulation can create a shock hazard due to a ground fault in the windings.

_____ 27. *True or False?* Megohmmeters are used to detect insulation failure along conductors.

17.2 Troubleshooting Electrical Motors

_____ 28. A good way to determine the condition of a hermetic compressor motor is to measure the unit's _____.

    A. power consumption

    B. total resistance

    C. capacitance

    D. None of the above.

_____ 29. Which of the following statements is true about capacitors?

    A. There are only three types of capacitors.

    B. One capacitor is as good as the next for any HVACR application.

    C. Capacitors use an expanded dimple to indicate when they have overloaded.

    D. None of the above.

_____ 30. A(n) _____ resistor slowly discharges a motor capacitor during the motor's Off cycle.

    A. bleed

    B. overload

    C. run

    D. start

_____ 31. *True or False?* A capacitor must be discharged before it can be tested with an ohmmeter.

_____ 32. *True or False?* Run capacitors are used intermittently for motor start-up.

_____ 33. Run capacitors are filled with _____ to dissipate large amounts of heat.

    A. low-pressure flowing nitrogen

    B. oil

    C. refrigerant

    D. water

_____ 34. The best way to protect a three-phase motor from the dangers of single-phasing is to install a(n) _____.

    A. infrared thermometer

    B. phase loss monitor

    C. power factor meter

    D. three run capacitors

6

17.3 Servicing Hermetic Compressor Motors

_____ 35. When checking a hermetic compressor motor, a technician can use an ohmmeter to _____.
   A. check the motor windings for continuity
   B. determine which terminal is which (if unmarked)
   C. measure the resistance between terminals
   D. All of the above.

_____ 36. The highest resistance value between two terminals on a single-phase hermetic compressor is between _____.
   A. C and R
   B. C and S
   C. R and S
   D. None of the above.

_____ 37. The highest resistance value between two terminals on a single-phase hermetic compressor is between _____.
   A. C and R
   B. C and S
   C. R and S
   D. None of the above.

_____ 38. If a short circuit develops in a hermetic compressor's motor, then _____.
   A. current draw decreases
   B. current draw increases
   C. the motor runs cooler than normal
   D. None of the above.

_____ 39. A common method of starting a stuck compressor (that contains a single-phase motor) is to _____.
   A. gently warm the entire dome using a blow torch
   B. install a hard start kit
   C. pour buckets of cold water over it
   D. solidly rap on the compressor dome with a rubber mallet

17.4 Servicing Fan Motors

_____ 40. Loose electrical connections create excessive resistance and can cause a fan motor to _____.
   A. hum loudly
   B. lose speed
   C. overheat
   D. All of the above.

_____ 41. Which of the following is _not_ a strategy that can be used when troubleshooting an ECM?
   A. Check for burnt terminals, loose connections, and frayed or broken wires.
   B. Open the control board and use a meter to measure the voltage supplied to the motor from the control board.
   C. Replace the motor controller board if it is found to be at fault.
   D. Check the low-voltage control circuit to make sure the motor is communicating with the HVACR system.

Name _____

_____ 42. *True or False?* If the blades of a fan have been badly forced out of position, spend at least an hour manually bending them back into shape and eyeballing the movement after each reposition.

17.5 Servicing External Motors

_____ 43. *True or False?* If a single-phase motor is humming but not turning, it means that electrical power is not reaching the motor.

_____ 44. When a single-phase motor's centrifugal switch has burnt and pitted contacts, the best option is to _____.
  A. coat the contacts with a dielectric spray
  B. file off the carbonized parts
  C. replace the contacts
  D. use sandpaper to smooth the surfaces of the contacts

_____ 45. When grease is used to lubricate a motor, it is normally _____.
  A. brushed on using any paintbrush
  B. poured in with an ordinary funnel
  C. pumped in by a grease gun into a zerk fitting
  D. smeared on by hand

_____ 46. *True or False?* A poorly aligned belt will shorten the life of a motor by causing it to continuously operate under an excessive load.

17.6 Servicing Motor Control Systems

_____ 47. *True or False?* If a heating or cooling system is not working, the thermostat may be at fault.

_____ 48. The best way to handle corroded contacts is to _____ them.
  A. file
  B. replace
  C. soak
  D. None of the above.

_____ 49. If an exact replacement for a potential relay is not available, _____ should be used.
  A. a current relay
  B. one of a higher voltage rating
  C. one of a lower voltage rating (90% of original)
  D. None of the above.

_____ 50. *True or False?* If a current relay is working properly, it should quickly close and then open again in about three seconds after power is applied to the motor circuit.

_____ 51. Potential relays that operate on counter emf from the motor require that the motor and the _____ operate correctly.

    A. thermostat

    B. capacitor

    C. overload cutout

    D. All of the above.

_____ 52. *True or False?* In general, motor-starting relays should be repaired instead of replaced.

Critical Thinking

53. Besides measuring specific resistances, list three or more electrical problems for which ohmmeters can be used to check.

_____

_____

_____

_____

_____

54. Why must only a single wire of a circuit be placed in the jaws of a clamp-on ammeter?

_____

_____

_____

55. What is the purpose of shunt resistors in an in-line ammeter?

_____

_____

_____

56. Before taking a voltage or current measurement, explain some of the checks you should do to prevent accidentally damaging the meter.

_____

_____

_____

_____

_____

_____

_____

Name _____

57. Explain why it is best to completely remove a device from its circuit when measuring its resistance.

_____

_____

_____

_____

_____

58. Briefly explain two ways that a technician can identify terminals on a hermetic compressor unit.

_____

_____

_____

_____

_____

59. List the two ways in which the pulley on a belt-driven unit must be in-line with a motor pulley.

_____

_____

_____

_____

6

Notes

Compressors

> *Carefully study the chapter and then answer the following questions.*

18.1 Compressor Drive Configurations

_____ 1. A(n) _____ prevents refrigerant leaks around the shaft of an open-drive compressor.
   A. compression ring
   B. cylinder head
   C. shaft seal
   D. valve plate

_____ 2. *True or False?* If the pulley on an open-drive compressor has a diameter three times greater than the pulley on the drive motor, the compressor shaft will rotate three times faster than the motor shaft.

_____ 3. *True or False?* A direct-drive compressor generally turns at a different speed than its motor.

_____ 4. *True or False?* The cases of fully hermetic compressors are typically bolted together.

_____ 5. Which of the following best describes the way fully hermetic compressors are lubricated?
   A. The mechanical systems are packed with grease and sealed at the factory and require no additional lubrication.
   B. They are lubricated by oil circulating with the refrigerant.
   C. They are lubricated by oil pumped in from an external reservoir.
   D. They are manually lubricated by technicians during seasonal inspections.

_____ 6. Compressor parts that are bolted together usually contain _____ to prevent leaks.
   A. cellophane
   B. gaskets
   C. grease
   D. refrigerant lubricant

18.2 Types of Compressors

Match each of the components listed with the corresponding letter in the following image.

_____ 7. Crankshaft

_____ 8. Crank throw

_____ 9. Connecting rod

_____ 10. Cylinder head

_____ 11. Piston

_____ 12. Valve plate

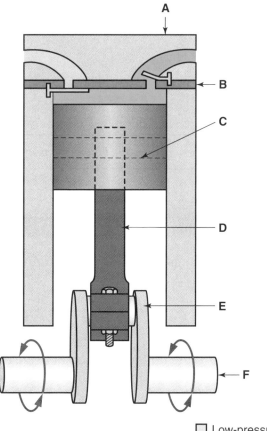

☐ Low-pressure vapor
■ High-pressure vapor

Goodheart-Willcox Publisher

Match each description with the compressor component it describes. Not all letters will be used.

_____ 13. Attaches the piston to the crankshaft

_____ 14. Moves up and down in the cylinder to draw in, compress, and discharge the refrigerant vapor

_____ 15. A shaft section that is larger and has a different center than the rest of the shaft

_____ 16. Connect pistons to connecting rods

_____ 17. Located between the cylinder head and cylinder block

_____ 18. Creates a seal between the piston and the cylinder wall

A. Connecting rod

B. Cylinder head

C. Eccentric

D. Piston

E. Piston pin

F. Piston ring

G. Valve plate

Name _____

_____ 19. *True or False?* The outer diameter of a piston ring is larger than the diameter of the cylinder.

_____ 20. A(n) _____ ring helps to lubricate the cylinder wall and prevent excess lubricant from entering the cylinder.

A. compression
B. eccentric
C. oil
D. shaft

_____ 21. The Scotch yoke type of reciprocating compressor does *not* use _____.

A. connecting rods
B. a crankshaft
C. cylinders
D. a piston

_____ 22. In rotating-vane compressors, the vanes may be brought into contact with the cylinder walls by spring pressure or by _____.

A. centrifugal force
B. discharge pressure
C. electromagnetism
D. gravity

Match the parts of a stationary-blade rotary compressor with the corresponding letters in the following illustration.

_____ 23. Roller

_____ 24. Blade

_____ 25. Cylinder

_____ 26. Rotor shaft

_____ 27. Housing

_____ 28. Eccentric

Goodheart-Willcox Publisher

_____ 29. *True or False?* In a scroll compressor, refrigerant vapor is compressed between two spinning scrolls.

_____ 30. In scroll compressors, refrigerant typically enters the scrolls from the _____ and exits out the _____.

A. center; sides
B. left side; right side
C. right side; left side
D. sides; center

_____ 31. *True or False?* Due to its method of compression, a screw compressor does not provide discharge pressure that is as smooth or pulsation-free as that of a reciprocating compressor.

_____ 32. In a centrifugal compressor, the spinning _____ flings refrigerant outward, compressing it against the casing.

   A. bearings
   B. eccentric
   C. impeller
   D. volute

18.3 General Compressor Components and Systems

_____ 33. Adding refrigerant and lubricant, checking pressure, and other procedures require the use of _____.

   A. mufflers
   B. O-rings
   C. service valves
   D. unloaders

_____ 34. An HVACR muffler uses a _____ to reduce noise caused by gas pulsations.

   A. helical passageway
   B. metal screen
   C. ribbed volute
   D. set of baffles

_____ 35. A measure of an oil's ability to flow at given temperatures is its _____.

   A. clearance
   B. eccentricity
   C. splash point
   D. viscosity

_____ 36. Which of the following statements about splash lubrication systems is _false_?

   A. The crankcase may be filled with oil up to the bottom of the main bearings.
   B. The crankcase may be filled with oil up to the middle of the crankshaft main bearings.
   C. Oil may be thrown around using scoops or dips.
   D. Clearances between moving parts in this type of system must be greater than in pressure systems.

_____ 37. An internal unloader is typically operated by _____.

   A. a direct acting solenoid
   B. a mechanical linkage connected to a centrifugal governor
   C. oil pressure controlled by a solenoid valve
   D. a stepper motor

_____ 38. To prevent liquid refrigerant slugging and oil slugging on compressors that operate in low ambient temperature, equip it with a(n) _____.

   A. crankcase heater
   B. oil pump
   C. O-ring
   D. unloader

Name _____

Critical Thinking

39. Assuming that the motors are single speed and not equipped with a VFD, explain why most belt-driven compressors turn at a different speed than their motors. How does this happen? How can compressor speed be increased? How can it be decreased?

    _____

    _____

    _____

    _____

    _____

    _____

40. List three conditions that are critical for proper belt-drive operation.

    _____

    _____

    _____

41. Why is a check valve installed in the discharge port of a rotary compressor?

    _____

    _____

42. List four benefits of scroll compressors over reciprocating compressors.

    _____

    _____

    _____

    _____

43. List the two major contributors to heat within a compressor.

    _____

    _____

44. List and explain three or more methods of cooling compressors.

    _____

    _____

    _____

    _____

    _____

Notes

Compressor Safety Components

Carefully study the chapter and then answer the following questions.

19.1 Compressor Operating Conditions

_____ 1. The primary purpose of refrigerant oil is to _____.
   A. cool the motor windings
   B. help refrigerant flow more easily
   C. lubricate the mechanical parts of a compressor
   D. prevent refrigerant from forming clumps

_____ 2. Why should refrigerant entering a compressor be all vapor and *not* liquid?
   A. Compressing liquid refrigerant would damage the refrigerant oil circulating with it.
   B. Liquid refrigerant carries too much heat content, so the system would reach its set point too quickly and short cycle.
   C. Liquid slugging can damage the compressor.
   D. None of the above.

_____ 3. If the moving parts within a compressor are not lubricated, it will most likely ____.
   A. create sparks that ignite the oil in the system
   B. friction weld the piston to cylinder
   C. have no short-term or long-term effect whatsoever
   D. produce small metal chips that will damage the system

19.2 Compressor Protection Devices

_____ 4. Devices that specifically provide overcurrent protection for compressors are _____.
   A. circuit breakers and fuses
   B. crankcase pressure regulators
   C. discharge line thermostats
   D. head pressure controls

_____   5. A thermal overload protects a compressor from overheating by opening _____.

   A. the compressor's power circuit
   B. a high-side relief valve
   C. a high-side service valve
   D. the liquid-line solenoid

_____   6. A crankcase pressure regulator protects a compressor from high pressure in _____.

   A. ambient air
   B. the conditioned space
   C. the high side of the system
   D. the low side of the system

_____   7. A discharge line pressure switch turns off the compressor if _____ gets too high.

   A. compressor temperature
   B. discharge line temperature
   C. head pressure
   D. liquid line temperature

_____   8. A discharge line thermostat turns off a compressor if _____ gets too high.

   A. condenser temperature
   B. discharge line temperature
   C. head pressure
   D. liquid line temperature

_____   9. The primary purpose of an accumulator is to protect a compressor from _____.

   A. high head pressure
   B. liquid refrigerant
   C. refrigerant oil
   D. vapor refrigerant

Name _____

19.3 Oil Control Systems

> *Match each of the oil control devices with the proper term.*

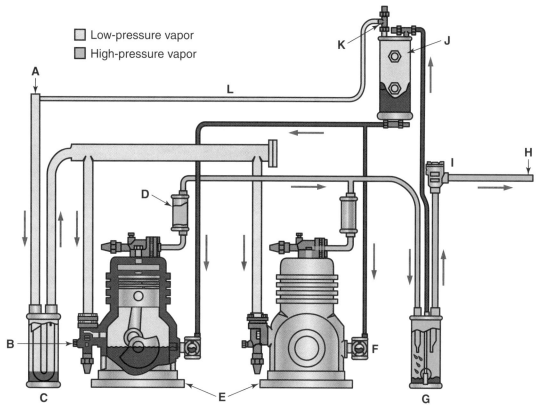

Low-pressure vapor
High-pressure vapor

Goodheart-Willcox Publisher

_____ 10. Accumulator

_____ 11. Check valve

_____ 12. Compressors

_____ 13. Compressor service valve

_____ 14. Discharge muffler

_____ 15. Oil level regulator

_____ 16. Oil reservoir

_____ 17. Oil separator

_____ 18. Pressure differential valve

_____ 19. Vent line

Match each of the components of the oil separator installation with the proper term. Not all letters will be used.

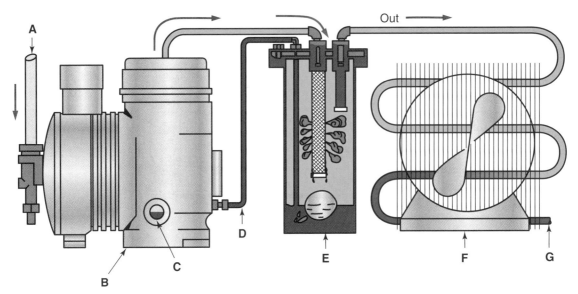

_____ 20. Liquid line

_____ 21. Oil level sight glass

_____ 22. Oil return line

_____ 23. Oil separator and reservoir

_____ 24. Suction line

_____ 25. *True or False?* An oil level regulator captures oil as it flows through screens, baffles, filters, or certain piping arrangements within it.

_____ 26. Which of the following statements regarding oil level regulators is *false?*

    A. Regulators are used in conjunction with oil separators and an oil return line to each compressor.

    B. The regulator allows oil to flow from the reservoir, through the regulator, and into the compressor.

    C. Only one oil level regulator is needed for multiple compressors piped in parallel.

    D. As the oil level drops in a compressor, its regulator returns oil to the compressor.

19.4 Vibration Absorbers

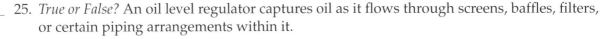

_____ 27. What attribute of vibration absorbers allows them to reduce vibrations?

    A. Absorptive

    B. Flexibility

    C. Permeability

    D. Rigidity

Name _____

_____ 28. To best eliminate vibrations, which part of a vibration absorber should be fastened to a wall?

    A. Middle

    B. Side closest to the compressor

    C. Side farthest from the compressor

    D. Any of the above.

_____ 29. *True or False?* Before installing a vibration absorber, stretch, twist, and compress it to prepare it for leakproof operation.

19.5 Crankcase Heaters

_____ 30. A crankcase heater is used to protect a compressor from _____.

    A. condensation forming on its shell

    B. excess oil slugging

    C. liquid refrigerant slugging

    D. vibration damage when cold weather solidifies its rubber grommets

_____ 31. A crankcase heater provides heat to a compressor from _____.

    A. discharge line refrigerant

    B. electric heating elements

    C. hot oil from the high side

    D. liquid line refrigerant

Critical Thinking

32. Explain what occurs when the oil level rises in the reservoir of an oil separator.

_____

_____

_____

_____

_____

33. Explain why a solenoid valve in the oil return line de-energizes and closes during the Off cycle. What might happen if the solenoid valve remained energized (open) during a long Off cycle?

_____

_____

_____

_____

7

34. What is the purpose of oil safety controls?

_____

_____

_____

_____

35. Briefly explain why vibration absorbers installed in both the suction and the discharge lines should be as close to the compressor as possible.

_____

_____

_____

36. Explain why the fittings and braided covering of a vibration absorber should *not* be overheated during installation. What might happen if they are overheated?

_____

_____

_____

_____

CHAPTER **20**

Metering Devices

Name _____

Date _____ Class _____

> *Carefully study the chapter and then answer the following questions.*

20.1 Metering Device Basics

_____ 1. The main purpose of a metering device's orifice is to produce _____.

A. equal distribution
B. a pressure drop
C. smooth flow
D. turbulent flow

_____ 2. Capillary tubes are used extensively in _____ systems.

A. chiller
B. motor vehicle
C. refrigerator and window ac unit
D. VFR

_____ 3. *True or False?* The limited amount of refrigerant allowed to pass through a metering device into the evaporator is at a high pressure.

_____ 4. *True or False?* The rate of flow through fixed metering devices is relatively constant when the compressor is running.

_____ 5. *True or False?* Because modulating metering devices can change refrigerant flow based on load, they operate more efficiently than fixed metering devices.

20.2 Capillary Tubes

_____ 6. For increased system efficiency, a capillary tube is usually positioned so that a portion of it is in contact with the _____.

A. compressor
B. condenser
C. liquid line
D. suction line

_____ 7. When capillary tube systems cycle off, _____.

    A. the high side goes low pressure, and the low side goes high pressure
    B. high side and low side pressures equalize
    C. the high side remains high pressure, and the low side remains low pressure
    D. None of the above.

_____ 8. *True or False?* The limited amount of refrigerant allowed to pass through a metering device into the evaporator is at a high pressure.

_____ 9. *True or False?* Extra refrigerant in a system using a capillary tube will cause the compressor motor to work harder during start-up.

_____ 10. An advantage of capillary tubes over modulating metering devices is that capillary tubes have no _____.

    A. adjustment screw
    B. flow change control
    C. load response mechanism
    D. moving parts

_____ 11. Capillary tube failure will most likely occur when it is _____.

    A. bent or crimped
    B. left undisturbed
    C. gently wiped clean
    D. None of the above.

_____ 12. When a capillary tube is replaced, it is good practice to also replace the _____.

    A. compressor
    B. condenser
    C. evaporator
    D. liquid line filter-drier

_____ 13. *True or False?* The refrigerant charge in systems using capillary tubes is critical.

20.3 Metering Orifices

_____ 14. *True or False?* Several fixed-orifice metering devices are often supplied by manufacturers for air conditioners that may need on-site changes to suction line length.

_____ 15. Piston-type metering orifices are most often used on _____ systems.

    A. chiller
    B. heat pump
    C. refrigerator-freezers
    D. window air conditioners

20.4 Thermostatic Expansion Valves (TXVs)

_____ 16. The main value that influences the operation of a TXV is _____.

    A. ambient temperature
    B. conditioned space temperature
    C. subcooling
    D. superheat

Name _____

_____ 17. If the pressure in a sensing bulb is _____ the combined pressure from the evaporator and the spring, a TXV's valve is forced open.

A. equal to
B. greater than
C. less than

_____ 18. When a thin layer of fluid adheres to a solid, but the two substances are not mixed together, _____ occurs.

A. absorption
B. adsorption
C. binding
D. initialization

_____ 19. When two substances are mixed and become one substance, _____ occurs.

A. absorption
B. adsorption
C. binding
D. initialization

_____ 20. The best superheat setting for an evaporator, called the minimum _____ signal (MSS) setting, is the point that causes the least sensing bulb temperature change while the system is running.

A. slugging
B. stable
C. starving
D. subcooling

_____ 21. *True or False?* The proper TXV and evaporator combination should keep surging and hunting to a minimum but still permit full evaporator use.

_____ 22. A TXV is operating properly when the evaporator is filled _____.

A. mostly with liquid
B. mostly with vapor
C. with equal parts liquid and vapor
D. Any of the above.

_____ 23. For a TXV adjusted to its MSS point, increasing spring tension on a TXV's adjustment screw will result in an increase of _____.

A. the amount of refrigerant entering the evaporator
B. the heat-absorbing capability of the evaporator
C. superheat
D. None of the above.

_____ 24. The condition in which liquid refrigerant is present in only part of the evaporator is called _____.

A. hunting
B. skimming
C. slugging
D. starving

_____ 25. *True or False?* By using an external equalizer, a TXV is able to provide the correct refrigerant flow, even when there are large pressure drops through the evaporator.

Match each of the following types of sensing bulbs to the statement that best describes it.

_____ 26. A sensing bulb charged with a refrigerant that is different from that in the system and which is completely vaporized at the desired temperature

_____ 27. A sensing bulb charged with a different refrigerant than what is in the system, and some of it is always in a liquid state

_____ 28. A sensing bulb that contains two substances: one is a noncondensing gas and the other is a solid

_____ 29. A sensing bulb charged with the same refrigerant as the system, some of which is always in a liquid state

_____ 30. A sensing bulb charged with the same refrigerant as the system and which is completely vaporized at a predetermined temperature

A. Adsorption gas cross-charged

B. Gas-charged

C. Gas cross-charged

D. Liquid-charged

E. Liquid cross-charged

_____ 31. On systems with small-diameter horizontal suction lines, the sensing bulb should be mounted on the _____ of the suction line.
A. bottom
B. side
C. top
D. Any of the above.

_____ 32. A sensing bulb must not be affected by the air or liquid being cooled. It should be wrapped in insulation so that only _____ affects the bulb.
A. head pressure
B. liquid line temperature
C. suction line temperature
D. suction pressure

_____ 33. Before a sensing bulb is installed, use steel wool to clean _____.
A. the sensing bulb
B. the sensing bulb and suction line
C. the suction line
D. None of the above.

_____ 34. *True or False?* Once all of the refrigerant in a gas cross-charged sensing bulb is vaporized, the valve is at its maximum operating pressure.

_____ 35. If the orifice of a TXV is oversized, too much refrigerant will pass into the evaporator and cause the suction line to _____ before the sensing bulb can close the valve.
A. burst
B. hunt
C. starve
D. sweat or frost

Name _____

_____ 36. If the superheat setting is increased to correct for an oversized TXV orifice, the evaporator will be _____ much of the time.

    A. empty

    B. flooded

    C. starved

    D. subcooled

_____ 37. The adjusting nut or screw on a thermostatic expansion valve is sensitive and should *not* be turned more than _____ turn while the unit is operating.

    A. one-sixteenth (1/16)

    B. one-quarter (1/4)

    C. one-half (1/2)

    D. one full

_____ 38. Some HVACR systems require special TXVs to accommodate _____.

    A. variations in operating conditions

    B. high system capacities

    C. multiple evaporators

    D. All of the above.

_____ 39. When a TXV is feeding refrigerant into an evaporator with multiple circuits, a(n) _____ should be installed.

    A. check valve

    B. distributor

    C. pressure limiter

    D. unloader

_____ 40. An HVACR system can be designed to use a lower torque compressor motor if the TXV is equipped with a(n) _____.

    A. check valve

    B. distributor

    C. pressure limiter

    D. unloader

_____ 41. A system equipped with a pilot-controlled thermostatic expansion valve is likely to be a _____.

    A. domestic refrigerator-freezer

    B. large, high-capacity system (50 tons and over)

    C. self-container water cooler

    D. small window air conditioner

20.5 Automatic Expansion Valves (AXVs)

_____ 42. *True or False?* Automatic expansion valve operation is based on its sensing bulb.

_____ 43. To maintain a constant pressure in the evaporator when the system is running, an AXV _____ when evaporator pressure begins to increase.

    A. closes completely

    B. decreases refrigerant flow

    C. increases refrigerant flow

    D. opens fully

_____ 44. The primary purpose of an AXV is to maintain constant _____.

    A. head pressure
    B. low-side pressure
    C. subcooling
    D. superheat

_____ 45. An AXV that is overcapacity may allow too much refrigerant into the evaporator and may cause condensation on the _____.

    A. compressor
    B. discharge line
    C. liquid line
    D. suction line

_____ 46. *True or False?* An expansion valve assembly with a faulty valve or seat could allow refrigerant to leak during the Off cycle and allow liquid refrigerant to flow into the suction line.

_____ 47. A compressor must start while under load if an automatic expansion valve does *not* have a(n) _____.

    A. bleeder or bypass
    B. distributor
    C. pilot valve
    D. pressure limiter

_____ 48. *True or False?* Since altitude can affect atmospheric pressure, an AXV needs to be adjusted to account for this.

_____ 49. If an AXV is undercapacity for its application, what will happen?

    A. The compressor will be in danger of liquid slugging.
    B. The evaporator will be flooded.
    C. The evaporator will be starved.
    D. Superheat will be low.

20.6 Electronic Expansion Valves (EEVs)

_____ 50. An EEV controls its orifice using a(n) _____.

    A. diaphragm
    B. electric operator
    C. float-operated needle valve
    D. spring assembly

_____ 51. During normal operation, many modern EEV controllers are able to use _____ as inputs to govern valve position and operation.

    A. pressure measurements
    B. temperature measurements
    C. temperature and pressure measurements
    D. only manual programming

_____ 52. A controller modulates the EEV through the use of an algorithm to maintain the desired _____.

    A. evaporator pressure
    B. head pressure
    C. subcooling
    D. superheat

Name _____

_____ 53. A PWM solenoid EEV controls its valve opening by _____.

    A. constant spring pressure

    B. holding the solenoid valve in an open position in discrete increments

    C. rapidly opening and closing its solenoid valve

    D. sensing bulb and diaphragm

20.7 Float-Operated Refrigerant Controls

_____ 54. A low-side float system maintains a constant level of liquid refrigerant in the _____.

    A. accumulator

    B. condenser

    C. evaporator

    D. liquid receiver

_____ 55. *True or False?* In a low-side float-operated control system, the liquid refrigerant is at a low temperature and will absorb considerable heat in both the On and the Off cycles.

_____ 56. *True or False?* The higher the temperature in a low-side float (LSF) flooded system, the lower the low-side pressure.

_____ 57. In evaporators equipped with a pan float design, the suction tube extends to the bottom of the pan to ensure that _____ collected in the pan is drawn into the suction line and recirculated.

    A. contaminant particulates

    B. liquid moisture

    C. liquid refrigerant

    D. oil

_____ 58. If a low-side float unit is not level, proper evaporation of refrigerant will not occur because a layer of _____ will cover the liquid refrigerant.

    A. acid

    B. ice

    C. oil

    D. water

_____ 59. *True or False?* It is only the float chamber, not the evaporator, that is flooded in a high-side float system.

_____ 60. In a high-side float system in which the evaporator is some distance from the float chamber, it is the _____ that controls the flow of refrigerant into the evaporator.

    A. AXV

    B. high-side float

    C. low-side float

    D. weight valve

_____ 61. A system with a high-side float uses a float chamber to perform the function of a(n) _____.

    A. accumulator

    B. condenser

    C. filter-drier and sight glass

    D. liquid receiver

_____ 62. *True or False?* In a high-side float system, when the liquid level on the high side drops, the valve opens.

Critical Thinking

63. List the three variables that affect capillary tube performance. Explain how each of these variables affects performance.

_____

_____

_____

64. What is the advantage of recent capillary tube designs having a larger diameter and being longer than older designs?

_____

_____

_____

_____

65. What is the purpose of a bleed port on a refrigerant metering device? What does this mean for the compressor?

_____

_____

_____

_____

66. Explain how the different types of TXV sensing bulbs are identified.

_____

_____

_____

_____

_____

67. How does a stepper motor operate an EEV's valve?

_____

_____

_____

_____

_____

Name _____

68. What is initialization? What is its purpose?

_____

_____

_____

_____

_____

69. How does the controller prevent the accumulation of missed steps in a system using an EEV?

_____

_____

_____

_____

70. List the different parts of a weight valve. What is a weight valve's purpose?

_____

_____

_____

_____

71. Explain why oil binding is less troublesome in high-side float systems than in low-side float systems.

_____

_____

_____

_____

Notes

Name _____

Date _____ Class _____

Carefully study the chapter and then answer the following questions.

21.1 Evaporators

_____ 1. An air-cooling evaporator _____.

    A. transfers heat from the air circulating over it to the refrigerant inside its tubing
    B. uses air as the primary medium for cooling the conditioned space
    C. conducts heat through the fins, to the tubing, then to the liquid refrigerant in the tubing
    D. All of the above.

_____ 2. For what reason does heat transfer somewhat slowly between air and the refrigerant in an air-cooling evaporator?

    A. Air has relatively low density.
    B. Heat does not conduct well between fins and tubing.
    C. Liquid refrigerant has low density.
    D. Refrigerant cannot absorb heat from metal very quickly.

_____ 3. Air defrosting is most commonly used on evaporator in _____ applications.

    A. cryogenic
    B. low-temperature
    C. medium-temperature
    D. high-temperature

_____ 4. A plain coil submerged and mounted inside a container filled with a liquid that provides good heat transfer is a(n) _____ evaporator.

    A. air-cooled
    B. forced-draft
    C. immersed
    D. natural-draft

_____ 5. Why might a designer of a liquid-cooling evaporator use a brine bath rather than a sweet water bath?

    A. Brine is less damaging to skin than sweet water.
    B. Brine baths require less maintenance than sweet water baths.
    C. So the bath could be cooled below 32°F (0°C) without freezing.
    D. To prevent the growth of bacteria.

_____ 6. Fin-and-tube evaporators circulating the most common types of refrigerants generally have fins made of ____.

    A. aluminum
    B. black iron
    C. stainless steel
    D. steel

_____ 7. In general, the more fins an air-cooling evaporator has, ____.

    A. the less its cooling ability
    B. the less its heat transfer
    C. the more its cooling ability
    D. All of the above.

_____ 8. Dead air and hot spots in a conditioned space can often be eliminated by using ____.

    A. baffles
    B. brine
    C. a condensate pump
    D. a reclaim tank

_____ 9. The air-cooling evaporators that are most effective at heat transfer are ____ evaporators.

    A. fin-and-tube
    B. microchannel
    C. plate
    D. All of the above.

_____ 10. Defrost timers commonly control the operation of the following, *except* for ____.

    A. electric heating elements
    B. expansion valves
    C. fan motors
    D. solenoid valves

_____ 11. Defrost timers end a defrost cycle using the following variables, *except* for ____.

    A. fluid flow
    B. pressure
    C. temperature
    D. time

_____ 12. The condition in which a defrost cycle continues too long and conditioned space temperature is raised too high is called ____.

    A. frosting the line
    B. hotboxing
    C. overdefrosting
    D. undercooling

_____ 13. When melted frost cannot be directly drained by gravity alone, it must be removed using a(n) ____.

    A. arrangement of eddy currents
    B. condensate pump
    C. pump-down solenoid
    D. set of baffles

Name _____

_____ 14. The method of defrost that involves circulating a hot liquid in and around the evaporator is _____ defrosting.
A. hot-gas
B. nonfreezing solution
C. pump-down
D. water

_____ 15. The method of defrost that involves circulating air from the conditioned space over an evaporator after the compressor cycles off is _____ defrosting.
A. hot-gas
B. electric
C. off-cycle
D. pump-down

_____ 16. The method of defrost that involves energizing electric heating elements is _____ defrosting.
A. hot-gas
B. electric
C. off-cycle
D. pump-down

_____ 17. The method of defrost that involves removing refrigerant from the evaporator is _____ defrosting.
A. hot-gas
B. electric
C. off-cycle
D. pump-down

7

_____ 18. The method of defrost that involves manually or automatically running tap water over an evaporator is _____ defrosting.
A. hot-gas
B. nonfreezing solution
C. pump-down
D. water

_____ 19. The method of defrost that involves circulating high-pressure refrigerant through an evaporator is _____ defrosting.
A. hot-gas
B. nonfreezing solution
C. pump-down
D. water

Match the component letters in this system equipped for two-pipe, reverse cycle hot-gas defrosting with the proper term. Not all letters will be used.

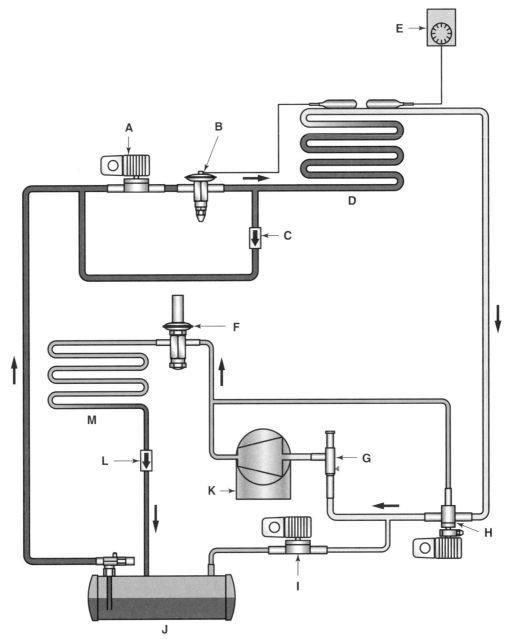

Goodheart-Willcox Publisher

_____ 20. Liquid line solenoid valve (open)

_____ 21. Defrost solenoid valve (closed)

_____ 22. Condenser check valve

_____ 23. Defrost thermostat

_____ 24. Crankcase pressure regulator

_____ 25. Pressure-regulating valve

_____ 26. Condenser

_____ 27. Three-way solenoid valve

_____ 28. Bypass check valve

_____ 29. Liquid receiver

Name _____

> *Examine the operational diagram of this reverse cycle hot-gas defrost for a multiple-evaporator system and determine whether the devices listed below are open (O) or closed (C). For each entry, enter either O for open or C for closed.*

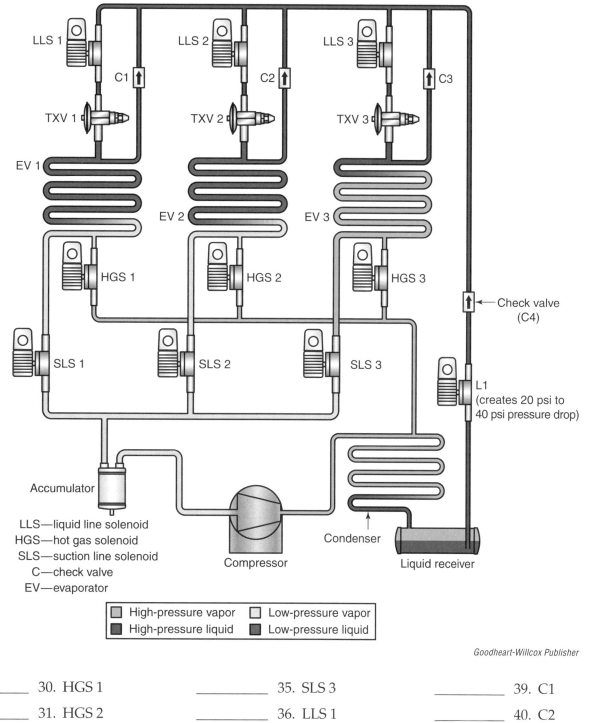

Goodheart-Willcox Publisher

_____ 30. HGS 1

_____ 31. HGS 2

_____ 32. HGS 3

_____ 33. SLS 1

_____ 34. SLS 2

_____ 35. SLS 3

_____ 36. LLS 1

_____ 37. LLS 2

_____ 38. LLS 3

_____ 39. C1

_____ 40. C2

_____ 41. C3

_____ 42. C4

_____ 43. *True or False?* When evaporator temperature cut-out point is reached in a system with pump-down defrost, the thermostat opens the circuit controlling a pump-down solenoid valve in the liquid line.

_____ 44. *True or False?* In a system with pump-down defrost, the evaporator fan continues to run while the system is in defrost mode.

21.2 Condensers

_____ 45. Refrigerant entering a condenser is a hot, high-pressure vapor, but refrigerant exiting a condenser should be a _____.
   A. cold, low-pressure liquid
   B. cool, low-pressure vapor
   C. hot, low-pressure vapor
   D. warm, high-pressure liquid

_____ 46. A condenser that is composed of a long refrigerant cylinder filled with straight copper tubes filled with cooling water is called a(n) _____ condenser.
   A. evaporator
   B. shell-and-coil
   C. shell-and-tube
   D. tube-within-a-tube

_____ 47. A condenser that is composed of a refrigerant cylinder filled with a winding spiral of copper tubing filled with cooling water is called a(n) _____ condenser.
   A. evaporator
   B. shell-and-coil
   C. shell-and-tube
   D. tube-within-a-tube

_____ 48. A condenser that is composed of a coaxial arrangement of tubing with refrigerant and water flowing in opposite directions is called a(n) _____ condenser.
   A. evaporator
   B. shell-and-coil
   C. shell-and-tube
   D. tube-within-a-tube

_____ 49. A condenser that uses fan drafts and water sprayed over its tubing is called a(n) _____ condenser.
   A. evaporator
   B. shell-and-coil
   C. shell-and-tube
   D. tube-within-a-tube

_____ 50. *True or False?* The majority of air-cooled condensers are located inside.

_____ 51. The most common build of air-cooled condenser is _____.
   A. bare coiled tube
   B. fin-and-tube
   C. microchannel
   D. plate

Name _____

21.3 Head Pressure Control

_____ 52. The reason for head pressure control in low ambient temperature can be explained by the concepts expressed in _____ law.

    A. Dalton's

    B. Gay-Lussac's

    C. Newton's

    D. Ohm's

_____ 53. The position of condenser air louvers for low ambient conditions is directly modulated by _____.

    A. ambient temperature

    B. crankcase pressure

    C. head pressure

    D. suction pressure

_____ 54. When head pressure control is done through fan cycling, hotter ambient temperature would result in _____.

    A. all condenser fans cycled off

    B. condenser fans turning more slowly

    C. fewer condenser fans operating

    D. more condenser fans in operation

21.4 Other Heat Exchangers

_____ 55. In a suction line-liquid line heat exchanger, system capacity is increased as the suction line experiences _____.

    A. less subcooling

    B. less superheating

    C. more subcooling

    D. more superheating

_____ 56. In a suction line-liquid line heat exchanger, system capacity is increased as the liquid line experiences _____.

    A. less subcooling

    B. less superheating

    C. more subcooling

    D. more superheating

_____ 57. In a suction line-liquid line heat exchanger, refrigerant generally flows _____.

    A. in opposite directions

    B. in the same direction

    C. Both A and B.

    D. Neither A nor B.

_____ 58. *True or False?* The higher the temperature in the liquid line, the greater the system's heat removal capacity.

7

_____ 59. Mechanical subcooling is generally used on _____.

    A. domestic refrigerator-freezers
    B. low-temperature commercial refrigeration systems
    C. small water coolers
    D. residential heat pumps

_____ 60. Sets of metal sheets with separate passageways sharing a significant amount of common surface area are assembled together to form _____.

    A. evaporative condensers
    B. microchannel heat exchangers
    C. plate heat exchangers
    D. shell-and-sheet condenser

Name _____

Match the components of the subcooler installation with the phrase that best describes it. Not all letters will be used.

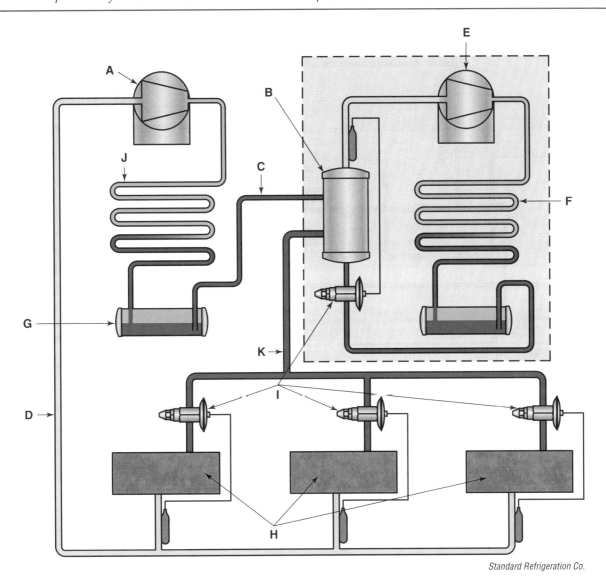

Standard Refrigeration Co.

7

_____ 61. Cooling cases

_____ 62. Subcooler

_____ 63. Subcooled liquid line

_____ 64. Warm liquid line

_____ 65. Liquid receiver

_____ 66. Suction line

_____ 67. Thermostatic expansion valves (TXVs)

Match each of the components on the series arranged heat recovery system with the proper term. Not all letters will be used.

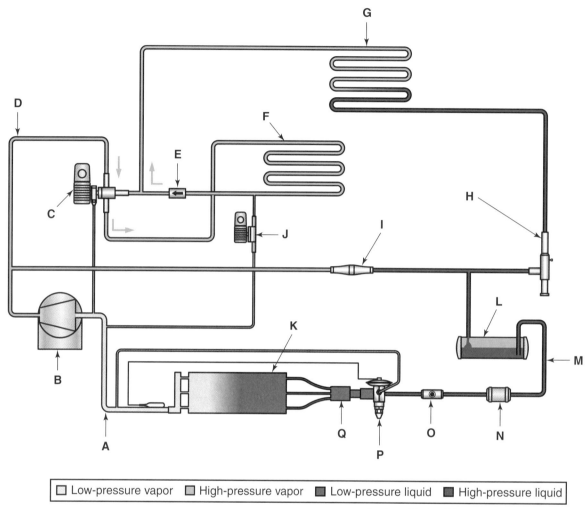

Courtesy of Sporlan Division – Parker Hannifin Corporation

_____ 68. Condenser

_____ 69. Discharge line

_____ 70. Condenser pressure regulator (ORI valve)

_____ 71. Suction line

_____ 72. Liquid receiver

_____ 73. Liquid line filter-drier

_____ 74. Three-way solenoid valve

_____ 75. Check valve

_____ 76. Sight glass

_____ 77. Receiver pressure regulator (ORD valve)

_____ 78. Liquid line

_____ 79. Solenoid valve

_____ 80. Distributor

_____ 81. Reclaim condenser

Name _____

_____ 82. Heat recovery systems most often use waste heat to warm _____.
  A. a building's hot water
  B. conditioned air in a building
  C. food products for meals
  D. public seating

Critical Thinking

83. Explain the effect that vapor bubbles and a film of oil coating the inside of evaporator tubing would have on system operation.

_____

_____

_____

_____

_____

84. Explain why heat must be applied to defrost evaporators used in low-temperature applications. Why can't system controls simply lock out normal system operation until the frost melts?

_____

_____

_____

_____

_____

85. Explain why and how gravitational circulation occurs in a natural-draft evaporator unit.

_____

_____

_____

_____

_____

86. Explain why it is better to have the evaporator coil located at the bottom of a soda fountain's sweet water bath, rather than at its top.

_____

_____

_____

_____

_____

87. Briefly explain how a water defrost system works.

_____

_____

_____

_____

_____

88. List at least three of the different methods of capacity control that can be used in a system with an evaporative condenser. Explain how capacity is controlled with each of these methods.

_____

_____

_____

_____

_____

89. Explain at least three of the reasons why air-cooled condensers are more commonly used in HVACR systems than water-cooled condensers.

_____

_____

_____

_____

90. Explain why outdoor air-cooled condensers used in year-round commercial refrigeration systems must be protected from strong winds and low ambient temperatures.

_____

_____

_____

_____

_____

_____

91. Explain the concept of the gas law that expresses the need for head pressure control for condensers operating in low ambient temperature.

_____

_____

_____

_____

_____

CHAPTER 22

Refrigerant Flow Components

Name _____

Date _____ Class _____

> *Carefully study the chapter and then answer the following questions.*

22.1 Refrigerant Loop Components

_____ 1. The four major components of a mechanical refrigeration system are _____.

    A. anti–short cycle controls, compressor, evaporator, and refrigerant metering device
    B. compressor, condenser, evaporator, and filter-drier
    C. compressor, condenser, defrost controls, and refrigerant metering device
    D. compressor, condenser, evaporator, and refrigerant metering device

_____ 2. Devices used for refrigerant flow control include the following, *except* _____.

    A. check valves
    B. liquid receiver
    C. pressure-regulating valve
    D. solenoid valves

_____ 3. A device used to maintain proper refrigerant quality is the _____.

    A. check valves
    B. filter-drier
    C. liquid receiver
    D. solenoid valves

22.2 Storage and Filtration Components

_____ 4. Install a _____ to protect a compressor after a burnout.

    A. liquid line solenoid valve
    B. liquid receiver
    C. sight glass
    D. suction line filter-drier

_____ 5. The main task of a drier is to _____.

    A. adsorb moisture and other contaminants that could mix to create acids
    B. baffle refrigerant to produce a more linear flow
    C. catch foreign materials circulating in the refrigerant
    D. show any bubbles flowing in the liquid line

_____ 6. *True or False?* Suction line filter-driers can hinder the return of oil to the compressor.

_____ 7. *True or False?* There should *not* be a measureable pressure difference in pressure measurements taken before and after the filter-drier.

_____ 8. A liquid receiver is a tank connected between the _____ that is used for refrigerant storage.

    A. compressor and condenser
    B. condenser and liquid line
    C. evaporator and suction line
    D. suction line and compressor

_____ 9. In a commercial refrigeration system that pumps down during the Off cycle, _____ controls the liquid line solenoid valve.

    A. cooling thermostat
    B. discharge line thermostat
    C. high-pressure control
    D. low-pressure control

_____ 10. Using Off-cycle pump-down on a commercial refrigeration system is a good method of ____.

    A. evaporator frost prevention
    B. head pressure control
    C. preventing compressor flooding
    D. short-cycle control

_____ 11. The sizing of a filter-drier is primarily based on a system's ____.

    A. capacity rating
    B. cost
    C. refrigerant type
    D. tubing material

_____ 12. *True or False?* A filter-drier should be replaced whenever the system is opened for service.

_____ 13. A small viewport installed in a refrigerant line describes a(n) ____.

    A. access port
    B. moisture indicator
    C. sight glass
    D. service valve

_____ 14. *True or False?* If a moisture indicator is hot, it may produce an inaccurate reading on the amount of moisture in a system.

Name _____

22.3 Refrigerant Flow Valves

_____ 15. *True or False?* Shutoff valves are made so that only a refrigeration service valve wrench can be used to turn them.

_____ 16. Systems that include manifold valves usually have multiple _____.

    A. condensers
    B. CPRs
    C. evaporators
    D. sight glasses

_____ 17. Refrigerant line valves _____.

    A. block refrigerant when flow reverses
    B. can be closed to cut off refrigerant flow through a section of piping
    C. restrict flow when preset exceeds a set limit
    D. All of the above.

_____ 18. A shutoff valve controlling refrigerant flow through a vertical length of refrigerant line is a _____ valve.

    A. check
    B. manifold
    C. riser
    D. service

_____ 19. A valve that allows fluid flow in only one direction is a _____ valve.

    A. check
    B. manifold
    C. riser
    D. service

_____ 20. *True or False?* The liquid line manifold in a multiple-evaporator system is equipped with hand valves.

_____ 21. *True or False?* The liquid line manifold distributes low-pressure liquid refrigerant to multiple evaporators.

_____ 22. The inefficient operation of a check valve may cause a(n) _____ noise.

    A. deep rumbling
    B. hammering
    C. humming
    D. scratching

7

Identify the parts of the solenoid valve shown in the following illustration.

_____ 23. Coil

_____ 24. Disc

_____ 25. Plunger

_____ 26. Plunger tube

_____ 27. Valve body

_____ 28. Valve cover

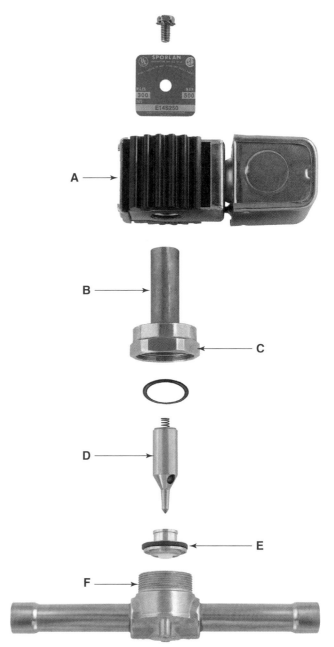

Courtesy of Sporlan Division – Parker Hannifin Corporation

Name _____

_____ 29. To prevent a TXV from opening intermittently during the Off cycle, a(n) _____ can most easily be used to automatically divert high-side pressure to the underside of the TXV's diaphragm during the Off cycle.

    A. check valve

    B. manifold valve

    C. service valve

    D. three-way solenoid valve

_____ 30. The pilot valve on a four-way solenoid valve _____.

    A. actuates the solenoid valve

    B. handles the majority of refrigerant flow

    C. provides a restriction for metered refrigerant flow

    D. uses high-pressure vapor to operate the four-way valve

> *The following illustrates a heat pump's reversing valve when the system is in heating mode. Match each of the letters with the proper tubing stub connection.*

_____ 31. Compressor discharge

_____ 32. Compressor suction

_____ 33. Indoor coil

_____ 34. Outdoor coil

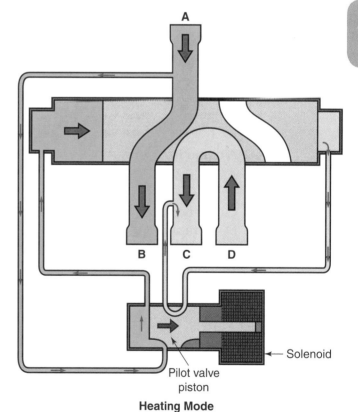

Heating Mode

_____ 35. A hot-gas defrost valve directs refrigerant into the low-side from the _____.

    A. condensate line
    B. discharge line
    C. liquid line
    D. None of the above.

_____ 36. A hot-gas bypass valve is primarily used for _____.

    A. alleviate high head pressures
    B. capacity control
    C. defrosting
    D. raise low head pressure

_____ 37. The two values on which hot-gas bypass valves operate is _____.

    A. head pressure and hot-gas temperature
    B. low-side pressure and head pressure
    C. low-side pressure and hot-gas temperature
    D. low-side pressure and low-side temperature

_____ 38. The value on which liquid injection valves operate is _____.

    A. evaporator temperature
    B. subcooling
    C. suction pressure
    D. superheat

_____ 39. A liquid injection valve is used in conjunction with a(n) _____.

    A. CPR
    B. EPR
    C. hot-gas bypass valve
    D. hot-gas defrost valve

_____ 40. A liquid injection valve's primary purpose is to _____.

    A. defrost evaporators
    B. raise suction pressure
    C. reduce excessive head pressure
    D. reduce the temperature of bypassed hot gas

Name _____

> *The following illustrates a refrigeration system with a method of capacity control. Match each of the letters of the components with the proper term.*

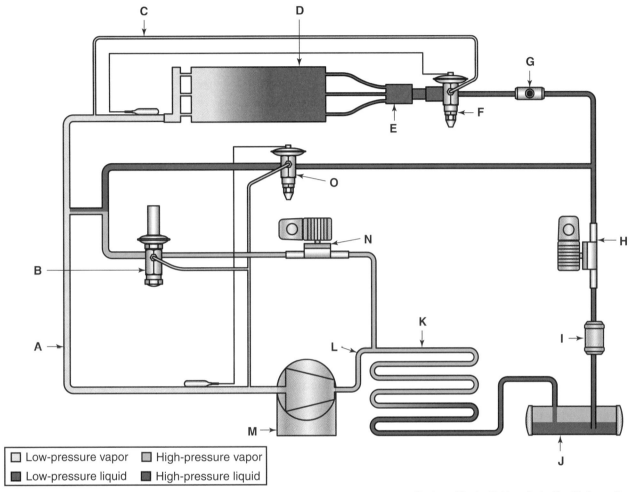

Courtesy of Sporlan Division – Parker Hannifin Corporation

_____ 41. Discharge line

_____ 42. External equalizer tubing

_____ 43. Hot-gas bypass valve

_____ 44. Hot-gas solenoid valve

_____ 45. Liquid injection valve

_____ 46. Liquid line filter-drier

_____ 47. Liquid line solenoid valve

_____ 48. Liquid receiver

_____ 49. Sight glass

_____ 50. Suction line

_____ 51. TXV

22.4 Pressure-Regulating Valves

_____ 52. As crankcase pressure rises, a crankcase pressure regulator _____.

A. bypasses
B. closes
C. opens
D. vents

_____ 53. The purpose of a CPR is to prevent _____.

A. compressor motor overload
B. high discharge temperature
C. high head pressure
D. low suction pressure

_____ 54. The calibration screw on a crankcase pressure regulator is adjusted according to which compressor rating?

A. Locked rotor amperage (LRA)
B. Rated compression ratio (RCR)
C. Rated horsepower (RHP)
D. Rated load amperage (RLA)

_____ 55. The primary purpose of evaporator pressure regulators (EPRs) on multiple-evaporator systems is to _____.

A. alleviate high head pressure
B. maintain different evaporator temperatures
C. prevent refrigerant migration
D. protect the compressor

_____ 56. EPRs are applied to refrigeration systems primarily based on the principle of _____ law.

A. Boyle's
B. Charles'
C. Dalton's
D. Gay-Lussac's

_____ 57. An electric evaporator pressure regulator differs from a standard EPR in that it uses a stepper motor to control its valve position and also measures and reacts to _____.

A. ambient temperature
B. evaporator pressure
C. evaporator temperature
D. suction pressure

_____ 58. Relief valves are generally used to protect against high _____.

A. ambient temperature
B. evaporator temperature
C. head pressure
D. suction pressure

Name _____

_____ 59. A relief valve with a metal core that melts at a low temperature is a ____.

    A. fusible plug

    B. rupture disc

    C. Schrader valve

    D. spring-loaded relief valve

_____ 60. A relief valve with an element that bursts before system pressure exceeds an unsafe level is a ____.

    A. fusible plug

    B. rupture disc

    C. Schrader valve

    D. spring-loaded relief valve

_____ 61. *True or False?* The pressure setting of a spring-loaded relief valve is easily adjusted in the field.

22.5 Head Pressure Control Valves

_____ 62. A condenser pressure regulator is a(n) ____ valve.

    A. close on rise of differential pressure

    B. close on rise of inlet pressure

    C. open on rise of differential pressure

    D. open on rise of inlet pressure

_____ 63. A receiver pressure regulator is a(n) ____ valve.

    A. close on rise of differential pressure

    B. close on rise of inlet pressure

    C. open on rise of differential pressure

    D. open on rise of inlet pressure

_____ 64. A refrigeration system with an outdoor air-cooled condenser that does not have head pressure control will likely suffer from ____.

    A. liquid slugging

    B. low or no subcooling

    C. relief valve venting

    D. starved evaporators

_____ 65. A low-ambient control valve is a combination of which two pressure regulators?

    A. Condenser and crankcase

    B. Condenser and evaporator

    C. Condenser and receiver

    D. Evaporator and crankcase

The following illustrates a refrigeration system with an outdoor, air-cooled condenser and one method of head pressure control. Match each of the letters of the components with the proper term.

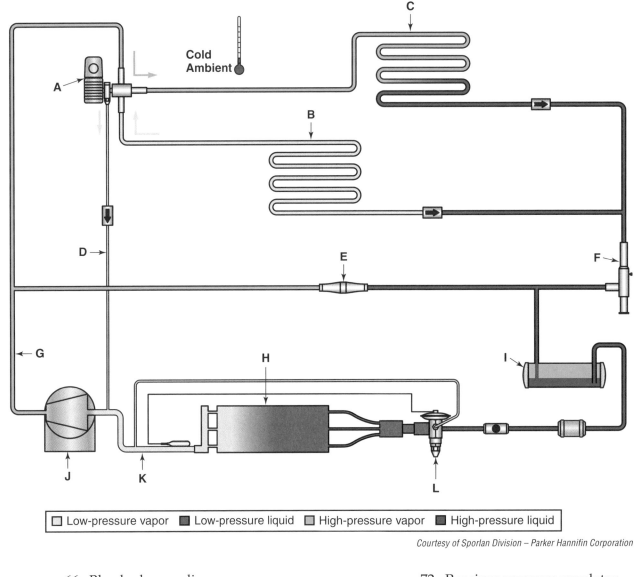

Low-pressure vapor   Low-pressure liquid   High-pressure vapor   High-pressure liquid

Courtesy of Sporlan Division – Parker Hannifin Corporation

_____ 66. Bleeder bypass line

_____ 67. Compressor

_____ 68. Condenser pressure regulator

_____ 69. Discharge line

_____ 70. Evaporator

_____ 71. Liquid receiver

_____ 72. Receiver pressure regulator

_____ 73. Split condenser valve

_____ 74. Suction line

_____ 75. Summer condenser

_____ 76. Summer/winter condenser

_____ 77. TXV

Name _____

Critical Thinking

78. Why should a filter-drier be installed in a relatively cool location?

_____

_____

_____

79. Explain why filter-driers in the suction line must be larger than those in the liquid line.

_____

_____

_____

_____

80. Explain why the outlet tubing of a liquid receiver starts near the bottom of the tank rather than the top.

_____

_____

_____

_____

81. List three indications or situation when a liquid line filter needs to be replaced.

_____

_____

_____

_____

82. Explain what may be indicated by bubbles in a sight glass. List and explain as many causes as you can recall.

_____

_____

_____

_____

83. Explain how refrigerant migration on the low side of a multiple-evaporator system occurs. How are check valves used in this situation?

_____

_____

_____

_____

_____

_____

_____

_____

84. Explain how check valves may be used in heat pumps.

_____

_____

_____

_____

_____

85. Explain why a hot-gas defrost valve may be needed in a refrigeration system.

_____

_____

_____

_____

86. If conditioned space temperature has not yet reached cut-out value but low-side pressure becomes excessively low, what does the hot-gas bypass valve do?

_____

_____

_____

87. If a liquid injection valve sensed either excessively low temperature or excessively low pressure, how would it respond? How would this response affect system operation? What would happen?

_____

_____

_____

_____

_____

Name _____

88. Explain the principles of the gas law that governs evaporator pressure regulator (EPR) operation. How is an EPR used? How does the gas law apply to its operation?

_____

_____

_____

_____

_____

_____

89. List three circumstances in which local codes usually require the installation of relief valves.

_____

_____

_____

_____

90. Explain why a relief valve is installed above the liquid level on a liquid receiver.

_____

_____

_____

91. Explain how a condenser pressure regulator and receiver pressure regulator work together to control head pressure. How do they operate? What do their actions accomplish?

_____

_____

_____

_____

_____

92. What is the purpose of a split condenser valve? What parts of a system does it affect? Explain all the functions of a split condenser valve. How is head pressure controlled using a split condenser valve?

_____

_____

_____

_____

_____

_____

Notes

CHAPTER **23**

Name _____

Date _____ Class _____

Overview of Domestic Refrigerators and Freezers

Carefully study the chapter and then answer the following questions.

23.1 Domestic Refrigeration

_____ 1. A benefit of refrigerating fruits and vegetables is that refrigeration _____ oxidation.

    A. enhances
    B. promotes
    C. slows
    D. speeds

_____ 2. Another benefit of refrigerating fruits and vegetables is that refrigeration _____ the multiplication of bacteria in the cells and fibers.

    A. increases
    B. promotes
    C. reduces
    D. speeds

_____ 3. Fruits and vegetables are prone to enzyme action and _____.

    A. blooming
    B. dehydration
    C. melting
    D. microbial growth

_____ 4. Meats and poultry are prone to spoilage from _____.

    A. blooming
    B. dehydration
    C. melting
    D. microbial growth

_____ 5. *True or False?* Enzymes are destroyed by fast freezing.

_____ 6. *True or False?* Dehydration in a freezer causes oxidation of food.

_____ 7. *True or False?* Food containers should be covered and airtight to keep foods moist and maintain a dry cabinet interior.

_____ 8. In what temperature range should the fresh food cabinet of a refrigerator be kept?

  A. −10°F to 0°F (−24°C to −17°C)
  B. 15°F to 20°F (−10°C to −7°C)
  C. 35°F to 40°F (2°C to 5°C)
  D. 45°F to 50°F (7°C to 10°C)

_____ 9. In what temperature range should the freezer cabinet of a refrigerator-freezer be kept?

  A. −10°F to 0°F (−24°C to −17°C)
  B. −50°F to −40°F (−46°C to −40°C)
  C. 35°F to 40°F (2°C to 5°C)
  D. 45°F to 50°F (7°C to 10°C)

23.2 Refrigerators and Freezers

_____ 10. A freezer that is accessed by a door on top is a(n) _____ freezer.

  A. chest
  B. hatchback
  C. overhead
  D. upright

_____ 11. A freezer that is accessed by a door on the side is a(n) _____ freezer.

  A. chest
  B. hatchback
  C. overhead
  D. upright

Match the parts of the freezer to their identifying phrases.

_____ 12. Compressor compartment

_____ 13. Condenser coil

_____ 14. Evaporator coil

_____ 15. Inner liner

_____ 16. Insulation

_____ 17. Outer shell

Goodheart-Willcox Publisher

Name _____

_____ 18. Which type of freezer loses the least amount of cool air when its door is open?

    A. chest
    B. hatchback
    C. overhead
    D. upright

_____ 19. The three main components of a frost-free freezer defrost system include the following, _except_ _____.

    A. defrost compressor
    B. defrost heater
    C. defrost thermostat
    D. defrost timer

_____ 20. A defrost timer in a frost-free freezer is wired so that it is either completing a circuit to the compressor or to the _____.

    A. defrost alarm
    B. defrost heater
    C. emergency trip switch
    D. None of the above.

_____ 21. A defrost heater is located close to the _____ coil and applies heat during the defrost cycle.

    A. compressor
    B. condenser
    C. evaporator
    D. None of the above.

_____ 22. The _____ monitors the temperature inside the cabinet and deactivates the heating coil if the temperature inside the cabinet rises above the set limit.

    A. defrost thermostat
    B. discharge line thermostat
    C. main thermostat
    D. thermal overloads

_____ 23. Condensate from defrosting is most often routed to _____.

    A. a drain line in the floor
    B. a drip pan
    C. the floor
    D. ice trays in the freezer

_____ 24. _True or False?_ A refrigerator-only unit may not include a defrost system because the warmer temperature inside the cabinet prevents frost from accumulating.

8

Match the following refrigerator and freezer terms with the phrase that best describes them.

_____ 25. Timer that counts down even when the compressor is not running.

_____ 26. The drying of food.

_____ 27. Rubber strip mounted to a refrigerator door.

_____ 28. Protein in food that triggers organic change.

_____ 29. Timer that counts down only when the compressor is running.

_____ 30. Drying of frozen food.

_____ 31. Removing frost and ice from a freezer by hand.

A. Continuous

B. Dehydration

C. Enzyme

D. Freezer burn

E. Gasket

F. Intermittent

G. Manual defrost

23.3 Innovative Technologies

_____ 32. The integration of media and refrigeration has resulted in higher end refrigerators having the following technology, *except* _____.

A. FM radio
B. self-propelled restocking capability
C. televisions
D. Wi-Fi enabled touch screens

_____ 33. Two-temperature compartment _____ coolers are now widely available for home use.

A. iced tea
B. lemonade
C. vodka
D. wine

Critical Thinking

34. Explain how freezer burn affects food.

_____

_____

35. Explain how a gasket forms an airtight seal between the door and refrigerator cabinet when the door is closed.

_____

_____

_____

_____

CHAPTER 24

Name _____

Date _____ Class _____

Systems and Components of Domestic Refrigerators and Freezers

> *Carefully study the chapter and then answer the following questions.*

24.1 Basic Components of Refrigerators and Freezers

_____ 1. Most compressors used in domestic refrigerators use _____ motors.

   A. ECM
   B. shaded-pole
   C. split-phase
   D. three-phase

_____ 2. *True or False?* The fuse or circuit breaker in an individual circuit should open in the event of continuous overload of more than 5%.

_____ 3. *True or False?* During startup, most motors used in domestic appliance compressors draw a current of only 60% of the running load.

_____ 4. A device wired in series with a compressor's motor windings that protects it from overload by increasing resistance with increasing temperature is a _____.

   A. circuit breaker
   B. fuse
   C. lockout relay
   D. PTC thermistor

_____ 5. *True or False?* The electrical cords for domestic refrigerators are insulated to withstand twice their normal voltage.

_____ 6. A condenser that uses tubes in direct contact with the outer wall of the cabinet so that the outer wall serves as a heat sink is a _____ condenser.

   A. forced-air
   B. hot-wall
   C. natural-convection
   D. subcooler

_____ 7. A forced-air condenser arrangement that takes air in along the front of the refrigerator and expels the air out of the back is commonly called a(n) _____ condenser.

   A. inshot
   B. roundabout
   C. single-pass
   D. two-pass

8

_____ 8. What type of metering device do most domestic refrigerators use?

    A. Automatic expansion valve

    B. Capillary tube

    C. Electronic expansion valve

    D. Thermostatic expansion valve

_____ 9. Domestic systems use a(n) _____ to prevent moisture from entering the metering device, where it could freeze and cause clogging.

    A. absorbant

    B. accumulator

    C. filter-drier

    D. moisture indicator

_____ 10. Heat is absorbed inside of the cabinet when refrigerant passes through the _____.

    A. accumulator

    B. condenser

    C. evaporator

    D. heat exchanger

_____ 11. What is the fixed temperature range commonly used for domestic refrigerator cabinets?

    A. 7°F to 11°F (–14°C to –12°C)

    B. 27°F to 31°F (–3°C to 0°C)

    C. 37°F to 41°F (3°C to 5°C)

    D. 57°F to 61°F (14°C to 16°C)

_____ 12. *True or False?* The temperature in a domestic refrigerator-freezer is controlled by a thermostat.

_____ 13. A thermostat that has a capillary tube that extends to a sensing bulb is a(n) _____ thermostat.

    A. electromechanical

    B. electronic

    C. magnetomotive

    D. pneumatic

_____ 14. A thermostat that provides direct sensing of a cabinet's interior temperatures using solid-state sensors is a(n) _____ thermostat.

    A. electromechanical

    B. electronic

    C. magnetomotive

    D. pneumatic

Name _____

24.2 Specialized Systems

_____ 15. The amount of air flowing between multiple compartments of a refrigerator-freezer is controlled by a _____.

A. air pump
B. damper
C. solenoid valve
D. weather vane

Match each of the descriptive phrases to the defrosting term it best describes. Not all letters will be used.

_____ 16. System in which the user presses a button to start the defrost cycle which automatically returns to regular operation after the unit is defrosted

_____ 17. An efficient system that determines the defrost intervals based on the amount of time the compressor runs

_____ 18. System that measures the time it takes to defrost the unit and then determines the interval before the next defrost cycle

_____ 19. A system that activates the defrost cycle based on the number of times the cabinet door is opened

_____ 20. A type of defrost timer that constantly counts down to the next defrost cycle, whether or not the compressor is running

_____ 21. System that produces no heat and that runs the evaporator fan when the compressor is not running

_____ 22. System that redirects hot, compressed vapor from the compressor through the evaporator to defrost it

_____ 23. A type of defrost timer that counts down to the next defrost cycle only when the compressor is running

A. Adaptive

B. Continuous

C. Cumulative run-time

D. Demand

E. Hot-gas

F. Intermittent

G. Off-cycle

H. Semiautomatic

8

Match the components of an electric defrost system with the corresponding letters in the following image.

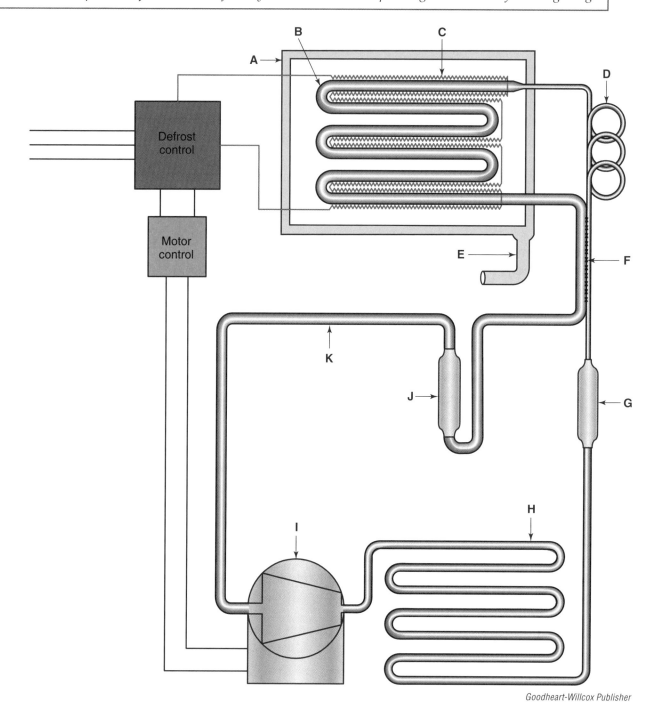

Goodheart-Willcox Publisher

_____ 24. Accumulator

_____ 25. Cabinet

_____ 26. Capillary tube

_____ 27. Compressor

_____ 28. Condenser

_____ 29. Drain

_____ 30. Evaporator

_____ 31. Filter-drier

_____ 32. Heat exchanger

_____ 33. Heating elements

_____ 34. Suction line

Name _____

Match the components of a hot-gas defrost system (during standard operation) with the corresponding letters in the following image.

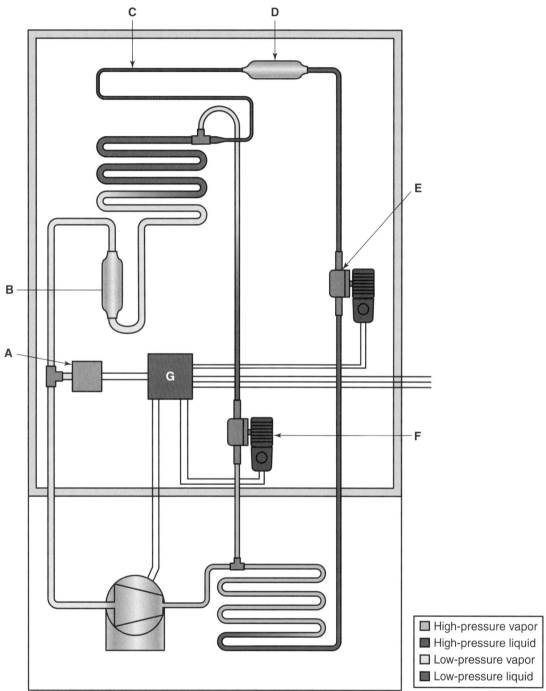

Goodheart-Willcox Publisher

High-pressure vapor
High-pressure liquid
Low-pressure vapor
Low-pressure liquid

8

_____ 35. Accumulator

_____ 36. Capillary tube

_____ 37. Defrost control

_____ 38. Filter-drier

_____ 39. Pressure
motor control

_____ 40. Solenoid
valve (closed)

_____ 41. Solenoid
valve (open)

_____ 42. Low-wattage electric heaters installed around the door openings to prevent condensation from forming are called _____ heaters.

A. condensation
B. defrost
C. mullion
D. post-condenser

_____ 43. Installed around the door openings, a _____ loop prevents condensation using the heat produced during the compression of the system's refrigerant.

A. condensation
B. defrost
C. mullion
D. post-condenser

_____ 44. *True or False?* Fruits are best kept at higher humidity levels than vegetables.

_____ 45. An automatic ice maker regulates the flow of water from the water supply into the ice mold using a _____.

A. damper
B. defrost condensation drip tube
C. solenoid valve
D. water pump

_____ 46. The three phases of operation of an automatic ice maker include the following, *except* _____.

A. dispense
B. fill
C. freeze
D. harvest

_____ 47. If an ice maker has a mold heater, it is attached _____ the ice mold.

A. above
B. away from the door side of
C. nearest the door side of
D. under

_____ 48. The level of harvested ice in an ice bin is continuously measured by the _____.

A. ejector
B. signal arm
C. thermostat
D. timing cam

_____ 49. An ice maker's ice is pushed out of its mold by the _____.

A. ejector
B. signal arm
C. thermostat
D. timing cam

_____ 50. Ice maker operation and devices are switched on and off by the _____.

A. ejector
B. signal arm
C. thermostat
D. timing cam

Name _____

_____ 51. A refrigerator-freezer's reservoir for its water dispenser is either a tank or a(n) _____.

    A. long coil of tubing in the refrigerated section

    B. a short single loop running through the freezer

    C. tube-within-a-tube heat exchanger

    D. All of the above.

Critical Thinking

52. Why is proper air circulation important with natural-draft condensers?

_____

_____

_____

_____

53. What purpose do cardboard partitions placed near a condenser serve?

_____

_____

_____

54. What two purposes does the heat exchanger between the capillary tube and suction line serve in a refrigerator-freezer unit?

_____

_____

_____

_____

55. What is the purpose of the slide control found on most crisper drawers? Why is this control adjustable?

_____

_____

_____

_____

_____

56. What could happen to an ice maker's ice mold if household water pressure is excessive?

_____

_____

_____

_____

8

57. What could happen if the household water pressure were excessively low?

_____

_____

_____

_____

58. Why is it important *not* to connect a refrigerator's water inlet to the outlet of a reverse osmosis filter?

_____

_____

_____

CHAPTER 25

Installation and Troubleshooting of Domestic Refrigerators and Freezers

> *Carefully study the chapter and then answer the following questions.*

25.1 Checking for Proper Installation

_____ 1. When a new refrigerator-freezer's compressor is mounted on or suspended from springs, the best time for a technician to remove the shipping bolts is ____.

   A. after installation
   B. at the factory
   C. before transportation from the store
   D. never

_____ 2. When a unit is located near a heat source, a(n) ____ should be installed on the side of the refrigerator-freezer that is next to the heat source.

   A. fan
   B. heat exchanger
   C. heat shield
   D. overload protection device

_____ 3. When leveling a refrigerator-freezer, use the following, *except* ____.

   A. adjustable supports
   B. multimeter
   C. spirit level
   D. wood shims

_____ 4. When installing a refrigerator with an automatic ice maker, less work is usually required to connect a(n) ____ valve mounted on the water line.

   A. automatic expansion
   B. saddle
   C. T-fitting and shut-off
   D. three-way solenoid

_____ 5. When a refrigerator-freezer has a hot-wall condenser, the unit should be at least ____ from the surrounding surfaces.

   A. 2″
   B. 6″
   C. 9″
   D. 12″

25.2 Diagnosing Symptoms

_____ 6. When the interior lights fail to turn on and off as the door opens and closes, a technician should first check for problems in the _____.

A. electrical circuit
B. high side of the system
C. ice maker
D. low side of the system

_____ 7. *True or False?* A leaky door gasket can cause a buildup of frost on an evaporator.

_____ 8. Sources of noise should be isolated using a(n) _____.

A. bubble solution
B. electronic leak detector
C. multimeter
D. ultrasonic leak detector

The following illustrates a test cord for a refrigeration system with a compressor that does not use any capacitors. Match each of the letters of the wiring terminals with the proper alligator clip.

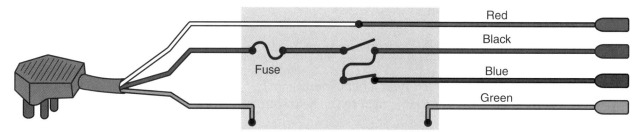

Goodheart-Willcox Publisher

_____ 9. Red   A. Common

_____ 10. Black   B. Ground

_____ 11. Blue   C. Run

_____ 12. Green   D. Start

25.3 Checking External Circuits

_____ 13. If a refrigerator-freezer's evaporator is icing over and remaining covered, a technician should first _____.

A. replace the capillary tube and filter-drier
B. replace the compressor
C. scrap the entire refrigerator-freezer
D. test the defrost timer

Name _____

_____ 14. When a refrigerator-freezer is not operating properly, all the following devices should be removed and tested outside the system, *except* the _____.

    A. compressor's capacitor

    B. condenser

    C. motor relay

    D. thermostat

_____ 15. A refrigerator-freezer's thermostat should always be connected into the _____ wire.

    A. common (grounded)

    B. ground

    C. hot (ungrounded)

    D. Any of the above.

_____ 16. Wiring that is used to ground appliances and electrical circuits should be _____.

    A. black

    B. green

    C. red

    D. white

_____ 17. A test cord is most properly used to test a refrigerator-freezer's _____ motor.

    A. two-pole

    B. three-phase

    C. four-pole

    D. Any of the above.

_____ 18. *True or False?* A two-pole motor should be tested by using only a properly sized relay in the circuit, else it may overheat.

_____ 19. When comparing open-circuit voltage to the voltage when the unit is running, the open-circuit voltage should be no more than _____.

    A. 5 V higher

    B. 5 V lower

    C. 25 V higher

    D. 25 V lower

_____ 20. The best instrument to use to check for continuity on mullion heaters is a(n) _____.

    A. in-line ammeter

    B. mullion meter

    C. test light or an ohmmeter

    D. voltmeter

_____ 21. If a refrigerator-freezer's thermostat is not working properly, the compressor _____.

    A. may keep running after the cut-out temperature is reached

    B. may not start when the cut-in temperature is reached

    C. starts when the thermostat is bypassed with a set of alligator clips

    D. All of the above.

25.4 Diagnosing Internal Troubles

_____ 22. In a capillary tube system, more refrigerant entering the evaporator than there is heat available to vaporize it may result from a(n) _____.

A. overcharge of refrigerant
B. system leak
C. totally blocked filter-drier
D. undercharge of refrigerant

_____ 23. Continuous running, no refrigeration, or a condenser that is cooler than normal may indicate a restriction in the following devices, *except* the _____.

A. capillary tube
B. filter-drier
C. screen on the high side
D. water solenoid valve

_____ 24. Rather than using piercing valves, an alternative method of connecting gauges and charging cylinders to a hermetic system is to use _____.

A. pinch pumps
B. punch holes
C. screw orifice
D. valve adapters

_____ 25. During the assembly process, a process tube is used for the following procedures, *except* _____.

A. charge
B. evacuate
C. flood
D. test

_____ 26. *True or False?* Corrosion problems can stem from the presence of moisture in a refrigerant circuit.

_____ 27. Moisture is prevented from circulating through a system and from reducing the chance of oil breakdown by the _____.

A. accumulator
B. compressor
C. defrost system
D. filter-drier

_____ 28. When a domestic appliance's metering device is clogged with wax, replace the _____.

A. filter-drier
B. metering device
C. filter-drier and metering device
D. entire appliance

_____ 29. When checking for a clogged capillary tube, listen for a _____ sound.

A. hissing
B. knocking
C. rumbling
D. screeching

Name _____

_____ 30. If an evaporator iced over because the electric defrost heating elements simply stopped producing heat and no other system operational symptoms were observed, a likely cause for the problem would be a(n) _____.

    A. electric short around the heating elements
    B. grounding of the heating elements
    C. mechanical binding in the compressor
    D. open in the heating element's circuit

Critical Thinking

31. Explain how to adjust the front supports of a refrigerator.

_____

_____

_____

_____

32. Why should an extension cord *not* be used between a refrigerator's electrical cord and an outlet?

_____

_____

33. Why should tubing running from the cold water line to a refrigerator-freezer include several large loops?

_____

_____

34. How can the circuit capacity (wire size, etc.) be checked at the moment of starting?

_____

_____

_____

_____

35. Explain why most refrigerator-freezers cannot restart compressor operation soon after it has cycled off.

_____

_____

_____

_____

36. Why is an inspection mirror useful to an HVACR technician? What can it be used to do?

_____

_____

_____

_____

37. Explain how a slip of paper is used to check for a faulty gasket.

_____

_____

_____

_____

38. List at least four common sources of unusual noises in a refrigerator.

_____

_____

_____

_____

_____

_____

_____

39. List at least three things that affect a refrigerator-freezer's cycling time and explain how they affect cycling time.

_____

_____

_____

_____

_____

40. Why is it useful to put matching marks on the fan hub and shaft before removing the fan from the motor shaft?

_____

_____

41. What might a partially frosted evaporator indicate?

_____

_____

_____

_____

_____

_____

Name _____

42. Explain how a technician can check pressures on a system that is *not* equipped with service valves.

_____

_____

43. Which values should a technician record when analyzing pressure-temperature conditions in a domestic appliance?

_____

_____

44. List five service tasks that are accomplished using service valves on hermetic systems.

_____

_____

_____

_____

_____

45. Why is it good practice to install a Schrader valve fitting if the system does not have one?

_____

_____

46. Explain why a refrigeration system with a sensing bulb thermostat calibrated for use at sea level would run too cold at elevations above 5000′.

_____

_____

_____

47. List some of the causes of wax separating from refrigeration oil.

_____

_____

_____

48. List some of the system symptoms that will happen if the amount of refrigerant in a capillary tube system is undercharged.

_____

_____

_____

_____

_____

49. What will happen if the amount of refrigerant in a capillary tube system is overcharged?

_____

_____

_____

_____

_____

_____

50. Explain what would result if the solenoid in a hot-gas defrost system failed with the valve in the open position.

_____

_____

_____

_____

CHAPTER **26**

Name _____

Date _____ Class _____

Service and Repair of Domestic Refrigerators and Freezers

> *Carefully study the chapter and then answer the following questions.*

26.1 External Service Operations

_____ 1. A system's compressor and condenser should be kept clean primarily because ____.
   A. cleanliness shows professionalism
   B. dirt and lint act as insulators to reduce heat transfer
   C. dirty components are shameful
   D. when a component is dirty, its warranty is voided

_____ 2. When cleaning components, dirt and lint should be blown off by the following, *except* ____.
   A. carbon dioxide
   B. nitrogen
   C. pressurized air
   D. refrigerant

_____ 3. When cleaning refrigerator-freezer parts with a vacuum cleaner, use a(n) ____ attachment.
   A. brush
   B. floor
   C. narrow nozzle
   D. scrubber

26.2 Internal Service Operations

_____ 4. To identify the different tubes connected to a compressor before replacing it, refer to the ____.
   A. electrical schematic
   B. manufacturer's diagram
   C. receipt of sale
   D. sales literature

_____ 5. When removing a hermetic compressor without using tubing cutters, the technician should _____.

A. clean the tubing or fittings at the compressor
B. put brazing flux on the connection
C. heat the joint and pull the tubing out of all the fittings
D. All of the above.

_____ 6. A compressor burnout involves the _____ in the compressor.

A. combustion of small amounts of refrigerant
B. ignition of small amounts of lubricant
C. deforming of the compressor discharge at high head pressure
D. melting or burning of internal components

_____ 7. Where can high head pressure occur in a domestic appliance?

A. At the accumulator's inlet
B. At the compressor's inlet
C. At the compressor's outlet
D. At the evaporator's inlet

_____ 8. A common cause of high head pressure is _____.

A. a clogged condenser
B. a clogged evaporator
C. cool ambient temperature around the appliance.
D. a low heat load inside the cabinet

_____ 9. _True or False?_ High temperatures created by high head pressure increase chemical action, leading to increased corrosion in the system.

_____ 10. When a burnt-out compressor is replaced, the following other components may need to be replaced, _except_ the _____.

A. filter-drier
B. hot-gas solenoid valve
C. metering device
D. suction line

_____ 11. If a compressor burnout is severe, the oil will be _____.

A. black and acidic with a pungent, very unpleasant odor
B. clear and brown with a mild odor of petroleum
C. thick and yellow with a smoky odor
D. white and foamy with a metallic odor

_____ 12. To remove paint and oxidation from the ends of suction, discharge, and oil stubs on a compressor, use 200-grit sandpaper or a(n) _____.

A. emery cloth
B. gas torch
C. hacksaw
D. single-cut file

_____ 13. If a replacement compressor's stub tubes are smaller in diameter than an appliance's suction and discharge lines before being swaged, the replacement compressor may _____.

A. be highly efficient
B. have insufficient capacity
C. have too much capacity
D. All of the above.

Name _____

_____ 14. After replacing a burned out compressor and a liquid line filter-drier, some technicians will install another filter-drier in the _____.

   A. between the capillary tube and evaporator
   B. discharge line
   C. liquid line
   D. suction line

_____ 15. Before applying epoxy to repair the leak in an aluminum evaporator, be sure that the evaporator is _____.

   A. at atmospheric pressure
   B. connected with low-pressure flowing nitrogen
   C. in vacuum
   D. pressurized to 30 psi at least

_____ 16. *True or False?* Brazing aluminum tubing should be avoided because it could weaken the tubing.

_____ 17. The best tool to use for properly cutting a capillary tube is a(n) _____.

   A. file
   B. hacksaw
   C. oxyacetylene torch
   D. tubing knife

_____ 18. The sealing caps on a new filter-drier should be removed _____.

   A. as soon as possible
   B. just before installation
   C. never
   D. only for pressure readings

_____ 19. The arrow stamped on the body of a filter-drier indicates the _____.

   A. direction in which refrigerant should flow
   B. first connection to make
   C. inlet of the filter-drier
   D. location of the compressor

_____ 20. Which of the following are signs that a hermetic system is low on refrigerant?

   A. Frequent cycling.
   B. Low head pressure.
   C. Low pressure on the low side.
   D. All of the above.

_____ 21. In cases where the recommended refrigerant amount is unknown, it may be necessary to charge a domestic appliance by observing _____.

   A. accumulator frosting
   B. condensation on the compressor
   C. filter-drier sweating
   D. frost back on the evaporator

_____ 22. *True or False?* It is often necessary to add oil to a hermetic system.

_____ 23. *True or False?* An overcharge of refrigerant oil will reduce the compressor's refrigerant-pumping capacity.

_____ 24. When using an ice and water solution to check the operation of a thermostat, the temperature of the ice and water solution must be stabilized at _____.

    A. 0°F (–17.7°C)
    B. 15°F (–9.5°C)
    C. 27°F (–2.7°C)
    D. 32°F (0°C)

_____ 25. Since freezer cabinet temperatures are usually in the 0°F to –20°F (–18°C to –29°C) range, use _____ instead of ice and water solutions to test a thermostat below freezing.

    A. brine solution under slight vacuum
    B. calibrating coolants
    C. dry ice
    D. superchilled solutions

Match each of the descriptive phrases to the term it applies to. Not all letters will be used.

_____ 26. The temperature at which a refrigeration system is activated.

_____ 27. The temperature at which a refrigeration system deactivates.

_____ 28. The group of temperatures that fall between cut-in and cut-out values.

_____ 29. The size of a temperature range.

_____ 30. A rating of a fluid's ability to flow at different temperatures.

A. Base line

B. Burnout

C. Cut-in temperature

D. Cut-out temperature

E. Fluidity

F. Starting point

G. Temperature differential

H. Temperature range

I. Viscosity

_____ 31. A chattering noise of a domestic appliance's thermostat is most likely caused by _____.

    A. a leaking sensing bulb
    B. pitted or burned contact points
    C. a short in the thermistor wiring
    D. stray emf

_____ 32. A plastic strip that connects a cabinet's outer shell to its liner is a _____.

    A. breaker strip
    B. liner fitting
    C. mullion
    D. shell strap

Name _____

26.3 Storing or Discarding a Refrigerator-Freezer

_____ 33. If a unit is being discarded, it is necessary to _____ the refrigerant inside the system.
  A. contaminate
  B. isolate
  C. reclaim
  D. vent

_____ 34. According to federal law, out-of-service refrigerator and freezer units must have their _____ removed.
  A. compressors
  B. condensers
  C. doors
  D. thermostats

Critical Thinking

35. What type of helpful information may be found on a system's identification plate?

_____

_____

_____

_____

_____

36. List three possible causes of compressor burnout.

_____

_____

_____

_____

_____

_____

37. If head pressure is too high, what steps can be taken to attempt to reduce the head pressure? Explain how each of these steps will help to reduce head pressure.

_____

_____

_____

_____

_____

8

38. How is the presence of hydrochloric and hydrofluoric acid in a refrigerant circuit detrimental to a refrigerator or freezer unit?

_____

_____

_____

_____

39. How is the oil in a system tested for acid when using an acid test kit?

_____

_____

_____

_____

_____

40. How can adjacent tubing and joints be protected during brazing?

_____

_____

41. Why is it important that a new replacement capillary tube have the same inside diameter (ID) and the same length as an old capillary tube being removed from a system?

_____

_____

_____

42. What is the best method for cutting a capillary tube? Explain how this is done.

_____

_____

43. Explain what undesirable result might occur to a capillary tube that is cut with a tubing cutter.

_____

_____

44. Explain why it is important to prevent a liquid line filter-drier from being exposed to too much heat.

_____

_____

45. Explain some of the advantages of installing a liquid line filter-drier just before the refrigerant metering device instead of being installed closer to the condenser.

_____

_____

_____

_____

_____

Name _____

46. Explain what it means that the refrigerant charge in small hermetic capillary tube systems is critical.

_____

_____

_____

_____

47. If a domestic appliance has a thermostat with a cut-in adjusting screw and a cut-out adjusting screw, what must the technician do to change the temperature range without changing the differential?

_____

_____

_____

_____

Notes

CHAPTER 27

Air Movement and Measurement

Name _____

Date _____ Class _____

Carefully study the chapter and then answer the following questions.

27.1 Climate

_____ 1. The conditions in the atmosphere refer to _____.
   A. air conditioning
   B. climate
   C. degree days
   D. weather

_____ 2. Long-term weather trends for a given region refers to _____.
   A. air conditioning
   B. atmosphere
   C. climate
   D. degree days

_____ 3. How warm the combination of humidity and temperature feels to an occupant of a space is _____ temperature.
   A. ambient
   B. climate
   C. equivalent
   D. relative

27.2 Atmosphere and Air

_____ 4. The atmosphere is approximately 23% _____ by weight.
   A. argon
   B. hydrogen
   C. nitrogen
   D. oxygen

_____ 5. _____ makes up 0.03% to 0.04% of the atmosphere and is absorbed by growing plants.

    A. Argon

    B. Carbon dioxide

    B. Hydrogen

    D. Oxygen

_____ 6. The majority of the atmosphere is composed of _____.

    A. argon

    B. hydrogen

    C. nitrogen

    D. oxygen

Identify the layers of the atmosphere illustrated in the following image.

_____ 7. Exosphere

_____ 8. Mesosphere

_____ 9. Stratosphere

_____ 10. Thermosphere

_____ 11. Troposphere

Goodheart-Willcox Publisher

_____ 12. A gas that is very light and present in most fuels is _____.

    A. argon

    B. hydrogen

    C. nitrogen

    D. oxygen

Name _____

_____ 13. According to Bernoulli's principle, _____ the velocity of air decreases the pressure.
   A. decreasing
   B. halving
   C. increasing
   D. maintaining

_____ 14. Air has weight, volume, and _____.
   A. temperature
   B. heat conductivity
   C. specific heat
   D. All of the above.

_____ 15. One pound of air at standard conditions (14.7 psia [101 kPa], 69.8°F [21°C]) occupies _____ ft².
   A. 1.3341
   B. 13.341
   C. 133.41
   D. 1,334.1

_____ 16. By dividing the weight of air by the volume it occupies, you can calculate the _____ of the air.
   A. density
   B. heat content
   C. pressure
   D. velocity

_____ 17. Devices used for measuring high temperatures include the following, *except* _____.
   A. glass-stemmed thermometers
   B. pyrometers
   C. thermistor thermometers
   D. thermocouple thermometers

_____ 18. The amount of moisture in an air sample compared to the total amount of moisture the same sample would hold if it were completely saturated at the same temperature defines the term _____.
   A. dew point
   B. relative humidity
   C. saturation
   D. total dew

9

Match the following points and lines on the red water vapor saturation curve to the descriptive phrase that it best represents.

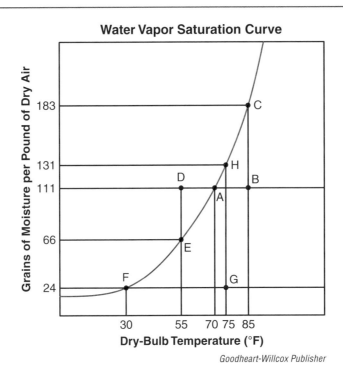

Water Vapor Saturation Curve

Goodheart-Willcox Publisher

_____ 19. A line showing how much additional moisture 85°F (29.4°C) air could hold.

_____ 20. A line representing an increase of only dry-bulb temperature.

_____ 21. A point showing a saturated condition with the most total grains of moisture per pound of air.

_____ 22. A point showing a saturated condition with the least total grains of moisture per pound of air.

_____ 23. A point containing 111 grains of moisture per pound of air at 85°F (29.4°C).

_____ 24. A line representing what happens when saturated air at 70°F (21.1°C) is cooled to a saturated temperature of 55°F (12.8°C).

A. Point A to B

B. Point B

C. Point C

D. Point A to D

E. Point A to E

F. Point F

G. Point B to C

_____ 25. _True or False?_ Since dry air prevents rapid evaporation of moisture from a surface, the surface does not cool as quickly as it would with moist air.

_____ 26. _True or False?_ The amount of moisture that a set volume of air can hold depends on air temperature.

_____ 27. An instrument that measures moisture in the air is a _____.
A. hygrometer
B. psychrometer
C. pyrometer
D. thermometer

Name _____

_____ 28. Desiccants are substances that absorb _____ from the air.

    A. moisture

    B. noncondensables

    C. noble gases

    D. solid particulates

_____ 29. In the cold season, humidity controls are typically set to _____ the conditioned air.

    A. add heat to

    B. add humidity to

    C. remove heat from

    D. remove humidity from

_____ 30. In the warm season, humidity controls are typically set to _____ the conditioned air.

    A. add heat to

    B. add humidity to

    C. remove heat from

    D. remove humidity from

_____ 31. *True or False?* Wet-bulb temperature will be higher than dry-bulb temperature at 100% humidity.

_____ 32. A(n) _____ chart is a graph of the properties of air.

    A. desiccant

    B. enthalpy

    C. hygrometric

    D. psychrometric

_____ 33. What do the red lines represent in the following chart?

    A. Constant dry-bulb temperatures

    B. Constant relative humidity

    C. Constant specific humidity levels

    D. Constant wet-bulb temperatures

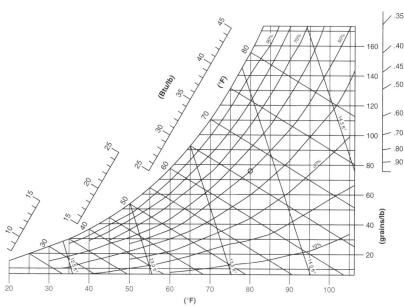

Goodheart-Willcox Publisher

_____ 34. What do the red lines represent in the following chart?

    A. Constant dry-bulb temperatures
    B. Constant specific humidity levels
    C. Constant specific volumes
    D. Constant wet-bulb temperatures

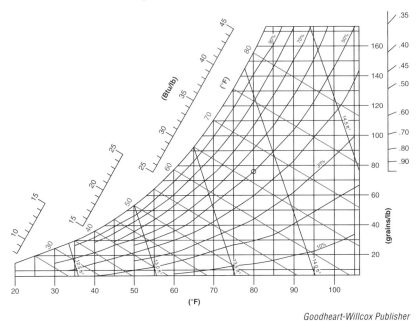

Goodheart-Willcox Publisher

_____ 35. What do the red lines represent in the following chart?

    A. Constant dry-bulb temperatures
    B. Constant specific humidity levels
    C. Constant specific volumes
    D. Constant wet-bulb temperatures

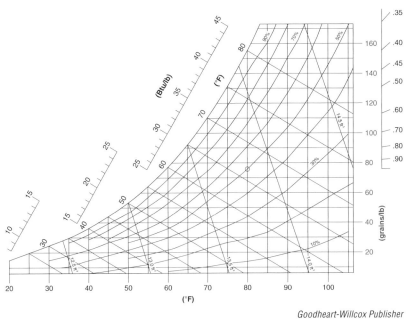

Goodheart-Willcox Publisher

_____ 36. *True or False?* The quantity of moisture air can hold decreases as its temperature increases.

Name _____

_____ 37. What do the red lines represent in the following chart?

    A. Constant relative humidity

    B. Constant specific humidity levels

    C. Constant specific volumes

    D. Constant total heat energy levels

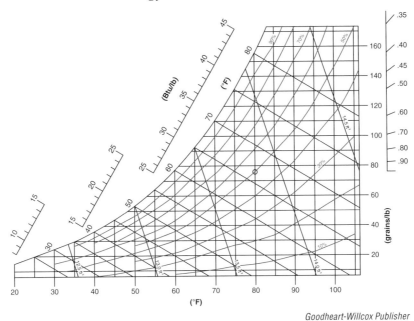

Goodheart-Willcox Publisher

_____ 38. What do the red lines represent in the following chart?

    A. Constant relative humidity

    B. Constant specific humidity levels

    C. Constant specific volumes

    D. Sensible heat ratios

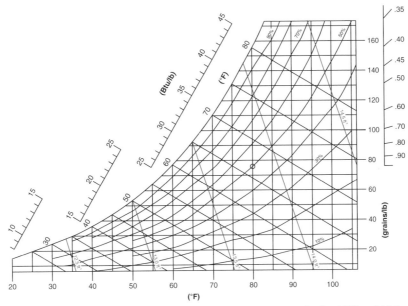

Goodheart-Willcox Publisher

_____ 39. Most people are comfortable in an atmosphere with the relative humidity between _____ and the temperature between 70°F and 85°F (21°C to 29°C).
   A. 0% and 70%
   B. 30% and 70%
   C. 70% and 90%
   D. 90% and 100%

27.3 Comfort Conditions

_____ 40. The combined effect of dry-bulb temperature, wet-bulb temperature, and air movement that provides an equal sensation of warmth and cold describes the term _____.
   A. degree day
   B. effective temperature
   C. psychrometry
   D. windchill index

_____ 41. In the summertime, air-conditioned buildings are usually kept at temperatures approximately _____ the outside temperature.
   A. 3°F to 6°F (2°C to 3.5°C) below
   B. 10°F to 15°F (5.6°C to 8.4°C) above
   C. 10°F to 15°F (5.6°C to 8.4°C) below
   D. the same as

_____ 42. _True or False?_ At comfortable temperatures, there is no sensation of either warmth or cold.

_____ 43. _True or False?_ In extreme heat, the body attempts to cool down by shutting down circulation to the extremities.

_____ 44. _True or False?_ In extreme cold, the body attempts to warm itself by increasing overall blood flow.

27.4 Air Movement

_____ 45. _True or False?_ An increase in wind velocity increases the heat loss of a heated structure, thereby impacting the efficiency of a heating system.

_____ 46. What forecast term on the Beaufort Scale is used to describe a wind velocity of 8–12 mph?
   A. Gale
   B. Gentle
   C. Hurricane-typhoon
   D. Storm

_____ 47. What forecast term on the Beaufort Scale is used to describe wind effects of breaking twigs off trees and generally impeding progress?
   A. Gale
   B. Gentle
   C. Hurricane-typhoon
   D. Light

_____ 48. _True or False?_ The calculated heat load for a structure should include the maximum wind velocity expected for the area.

Name _____

_____ 49. During the winter months, the _____ is used to describe the combined effect of temperature and wind speed.

   A. air velocity measurement
   B. Comfort Health Index
   C. sensible heat ratio
   D. windchill index

_____ 50. *True or False?* The exposure time required for frostbite to occur becomes shorter as wind speed decreases.

_____ 51. The volume of air flowing through a duct is calculated in _____.

   A. feet per minute
   B. cubic feet per minute
   C. miles per hour
   D. feet squared

_____ 52. An anemometer using current conducted through tungsten to calculate velocity is a _____ anemometer.

   A. hot-wire
   B. PTC
   C. swinging vane
   D. vane

_____ 53. An anemometer using incoming air to deflect a vane at different angles to calculate velocity is a _____ anemometer.

   A. hot-wire
   B. PTC
   C. swinging vane
   D. vane

_____ 54. An anemometer using a small propeller and electronic circuitry to calculate velocity is a _____ anemometer.

   A. hot-wire
   B. PTC
   C. swinging vane
   D. vane

_____ 55. If the air is in motion when a technician is measuring airflow using a pitot tube and manometer, the _____.

   A. static pressure is always greater than the total pressure
   B. fluid in the manometer is pushed toward the static pressure side
   C. pressures will equalize, resulting in no measurable value
   D. All of the above.

_____ 56. A pitot tube and manometer used for ductwork airflow in HVAC most often measure values in _____.

   A. inches of mercury (in. Hg)
   B. inches of water (in. H_2O)
   C. microns
   D. psi

_____ 57. *True or False?* Velocity pressure is exerted when the airflow is slowed or stopped.

9

_____ 58. *True or False?* With an anemometer that can be placed over an entire duct grille, a technician can calculate the number of Btus going into the space through each grille by multiplying the volume of airflow by the appropriate temperature factor.

_____ 59. *True or False?* To obtain correct velocity value, a technician should take several readings in various parts of the duct and then average the readings.

_____ 60. A pitot tube should be used only where the duct is very _____.
A. curved
B. long
C. round
D. short

_____ 61. The process of cold air drops and warm air rises resulting in a gradual difference in air temperature from floor to ceiling is known as _____.
A. climatizing
B. stratification
C. total heat indexing
D. ventilation

_____ 62. Make-up air is necessary for fuel-burning appliances because oxygen is necessary for _____.
A. climatizing
B. combustion
C. stratification
C. ventilation

27.5 Factors Affecting Indoor Air Conditions

_____ 63. *True or False?* Black absorbs more heat than white.

_____ 64. *True or False?* Yellow absorbs more heat than red.

_____ 65. *True or False?* Chrome does not absorb heat easily.

_____ 66. *True or False?* Polished metal does not radiate heat efficiently.

_____ 67. *True or False?* Because glass is a rather poor conductor of heat, greenhouses and other buildings with a lot of glass rarely get hot on sunny days.

_____ 68. When heat rays strike a surface of the same _____ as the radiating body, there is no increase or decrease of heat in the body being struck by the radiation.
A. color
B. shape
C. size
D. temperature

_____ 69. *True or False?* Vapor barriers should be installed on the cold side of a heated space.

_____ 70. *True or False?* Vapor barriers prevent water vapor from passing from warm to cold surfaces.

_____ 71. Water vapor condenses into droplets when its temperature drops to the _____.
A. degree day average
B. dew point
C. effective temperature
D. equivalent temperature

Name _____

_____ 72. Reducing heat conductivity with _____ helps maintain desired air-conditioning
temperatures more economically.
   A. heat sinks
   B. insulation
   C. psychrometric paper
   D. vapor barriers

For questions 73–76, use the following chart to answer the questions.

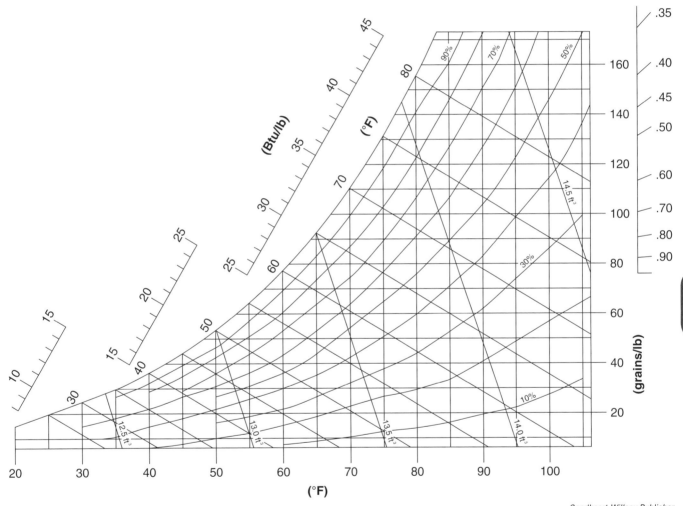

_____ 73. A sample of air has a dry-bulb temperature of 90°F (32°C) and a relative humidity of 60%.
Determine the wet-bulb temperature.
   A. 34°F
   B. 59°F
   C. 64°F
   D. 77°F

_____ 74. A sample of air has a dry-bulb temperature of 80°F (27°C) and a relative humidity of 40%. Determine the wet-bulb temperature.

    A. 25°F

    B. 38°F

    C. 57°F

    D. 63°F

_____ 75. Find the relative humidity when the dry-bulb temperature is 65°F (18°C) and the specific humidity is 70 grains per pound of dry air.

    A. 18%

    B. 68%

    C. 76%

    D. 100%

_____ 76. Find the relative humidity when the dry-bulb temperature is 85°F (29°C) and the specific humidity is 100 grains per pound of dry air.

    A. 11%

    B. 33%

    C. 57%

    D. 100%

Critical Thinking

77. In your own words, summarize the ASHRAE definition of _air conditioning_. Be sure to include all the aspects involved in the concept.

_____

_____

_____

_____

_____

_____

78. Compute the heating degree days with the following information. The low temperature for a certain day was 42°F and the high temperature for the same day was 60°F. Remember that degree day calculations are based on 65°F (18°C). Show your calculations.

_____

Name _____

79. Explain why the red lines in the following chart all end at different points along the left side of the graph. What does this mean concerning the properties of air?

_____

_____

_____

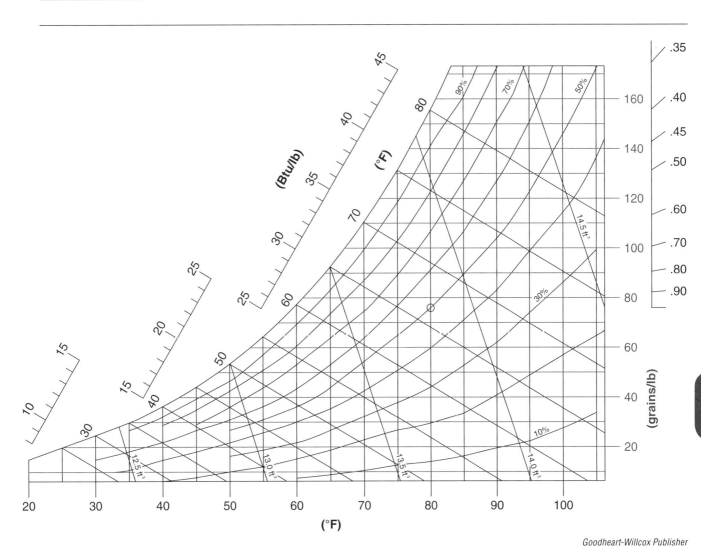

Goodheart-Willcox Publisher

80. Explain how a technician would plot the point for a sample of air on a psychrometric chart when the values for enthalpy and sensible heat ratio are available. Also, how would the other values be determined?

_____

_____

_____

_____

Notes

CHAPTER **28**

Air Quality

Name _____

Date _____ Class _____

28.1 Indoor Air Quality Standards and Guidelines

_____ 1. ASHRAE Standard 62 covers _____.
   A. equipment requirements
   B. system requirements
   C. minimum ventilation rates
   D. All of the above.

_____ 2. Standard 62.1 ties ventilation rates to the number of building _____ in a given space.
   A. doors
   B. machines
   C. occupants
   D. windows

28.2 Air Pollutants

_____ 3. A substance that is in some way a detriment to comfort, health, and a desirable environment is called a(n) _____.
   A. danger
   B. hazard
   C. OSHA violation
   D. pollutant

_____ 4. To what agency did the Federal Clean Air Act of 1970 give the power to establish and enforce standards for clean air?
   A. Center for Disease Control and Prevention
   B. Environmental Protection Agency
   C. Public Health Service
   D. Risk Management Agency

9

_____   5. Small particles that become temporarily airborne from wind, a sudden earth disturbance, or mechanical work on a solid are referred to as _____.

   A. bioaerosols
   B. dust
   C. fumes
   D. smoke

_____   6. Small particulates suspended in the air formed by the condensation and solidification of gaseous materials is referred to as _____.

   A. bioaerosols
   B. dust
   C. fumes
   D. smoke

_____   7. The term for solid particles (0.1–13 microns) carried into the atmosphere by gaseous products of incomplete combustion is _____.

   A. bioaerosols
   B. dust
   C. fumes
   D. smoke

_____   8. The term for simple, single-cell microorganisms that are responsible for many diseases is _____.

   A. asbestos
   B. bacteria
   C. mold
   D. pollen

_____   9. The term for minute fungi whose spores may be able to cause illness is _____.

   A. asbestos
   B. bacteria
   C. mold
   D. pollen

_____   10. The term for small particles released by plants as part of their reproductive cycle is _____.

   A. asbestos
   B. bacteria
   C. mold
   D. pollen

_____   11. The naturally occurring mineral that is now a known carcinogen but was formerly used widely in construction and as insulation is _____.

   A. asbestos
   B. bacteria
   C. mold
   D. pollen

_____   12. An odorless, colorless gas that replaces oxygen in red blood cells and is produced by incomplete combustion of fuel describes _____.

   A. carbon dioxide (CO_2)
   B. carbon monoxide (CO)
   C. nitrogen oxide (NO_x)
   D. sulfur dioxide (SO_2)

Name _____

_____ 13. A common gaseous pollutant produced by burning coal, gas, or oil that may contain carcinogens describes _____.
- A. carbon dioxide (CO_2)
- B. ozone (O_3)
- C. radon (Rn)
- D. sulfur dioxide (SO_2)

_____ 14. An odorless, tasteless, radioactive, inert gas in soil and rocks that can accumulate in homes through cracks describes _____.
- A. carbon dioxide (CO_2)
- B. ozone (O_3)
- C. radon (Rn)
- D. sulfur dioxide (SO_2)

_____ 15. A gas that is produced photochemically, acts as a disinfectant, and removes odors but is harmful in large concentrations describes _____.
- A. carbon dioxide (CO_2)
- B. carbon monoxide (CO)
- C. ozone (O_3)
- D. radon (Rn)

_____ 16. A nontoxic gas that tends to build up in spaces with poor ventilation and can cause fatigue, headaches, and discomfort in higher concentrations describes _____.
- A. carbon dioxide (CO_2)
- B. carbon monoxide (CO)
- C. nitrogen oxide (NO_x)
- D. radon (Rn)

_____ 17. An ozone concentration of _____ parts per million (ppm) is generally considered the maximum level permissible for an eight-hour exposure.
- A. 0.1
- B. 1.0
- C. 10
- D. 100

9

28.3 Indoor Air Quality

_____ 18. *True or False?* Removing lint around and under a clothes dryer on a regular basis can improve the quality of air inside a home.

_____ 19. In order to maintain adequate ventilation and filtration, a building's HVAC system must be designed in accordance with ASHRAE standards _____.
- A. 17 and 32.1
- B. 62.1 and 129
- C. 90.1 and 188
- D. 90.2 and 189.1

_____ 20. When bioaerosols and airborne particulates are the source of IAQ problems, the best solution may be to install _____.

    A. a better thermostat

    B. a germicidal light

    C. an HRV or ERV

    D. a humidifier and dehumidifier

_____ 21. When volatile organic compounds (VOCs) are the source of IAQ problems, the best solution may be to install _____.

    A. a better thermostat

    B. a germicidal light

    C. an HRV or ERV

    D. a humidifier and dehumidifier

_____ 22. When moisture and increased fungal growth are the source of IAQ problems, the best solution may be to install _____.

    A. a better thermostat

    B. a germicidal light

    C. an HRV or ERV

    D. a humidifier and dehumidifier

_____ 23. When stale air and too much carbon dioxide are the source of IAQ problems, the best solution may be to _____.

    A. install an air purification system

    B. install an HRV or ERV

    C. install a humidifier and dehumidifier

    D. repair the furnace or gas appliance

_____ 24. When too much carbon monoxide is the source of IAQ problems, the best solution may be to _____.

    A. install an air purification system

    B. install an HRV or ERV

    C. install a humidifier and dehumidifier

    D. repair the furnace or gas appliance

_____ 25. Sick building syndrome can usually be confirmed when at least _____% or more of a building's occupants complain of drowsiness, fatigue, eye and skin irritations, or respiratory problems.

    A. 5

    B. 20

    C. 50

    D. 80

_____ 26. Multiple chemical sensitivity (MCS) is a condition of sensitivities and adverse reactions to low levels of _____.

    A. bacteria, viruses, and mold

    B. bioaerosols, chemicals, and other irritants

    C. carbon dioxide, carbon monoxide, and ozone

    D. dust, fumes, and smoke

Name _____

28.4 Air Cleaning

_____ 27. Air filter efficiency is measured by the following three factors, *except* the _____.
  A. degree of discoloration on the exhaust side of the filter being tested
  B. size of the smallest particle that can be removed
  C. strength of the odor of a used filter
  D. total weight of dirt collected

_____ 28. Air filters and electronic air cleaners are rated according to ASHRAE Standard _____.
  A. 52.2
  B. 62.1
  C. 90.1
  D. 129

_____ 29. *True or False?* A manufacturer's filter efficiency rating usually refers to an air filter's ability to capture only large particles (3.0–10.0 microns).

> *Match each of the following types of air filters to the phrase that best describes them.*

_____ 30. Inexpensive panels made for single use.   A. Carbon

_____ 31. Designed to be used repeatedly and may be made of synthetic media.   B. Disposable

_____ 32. Uses an electrical charge on particles in the air.   C. Electrostatic

_____ 33. A type of filter known for its ability to remove odors.   D. HEPA

_____ 34. High-efficiency air filter that removes a high percentage of particles   E. Washable
       with an air purifier.

9

_____ 35. In most residential applications, disposable filters have a thickness of _____.
  A. 1/2″
  B. 1″
  C. 3″
  D. 5″

_____ 36. Disposable air filters are often made from _____.
  A. aluminum mesh
  B. fiberglass
  C. polyester and foam
  D. wire cloth

_____ 37. The activated charcoal in a _____ filter will adsorb as much as 50% of its weight in foreign gases.
  A. carbon
  B. disposable
  C. HEPA
  D. passive electrostatic

_____ 38. To qualify as a HEPA filter under US government standards, a filter must capture _____ of airborne particles 0.3 microns or larger.
   A. 87.97%
   B. 90.92%
   C. 99.97%
   D. 100%

_____ 39. Air filters should be changed at approximately _____% of their design final resistance.
   A. 10
   B. 25
   C. 80
   D. 100

_____ 40. A ventilation system is usually designed to allow the air filter pressure drop to be about _____ of the total pressure drop.
   A. one-fourth
   B. one-half
   C. two-thirds
   D. three-quarters

_____ 41. The measure of the ability of an air filter to remove atmospheric dust by evaluating the flow rates on both sides of the air filter and also the quantity of material it captures describes _____.
   A. atmospheric dust spot efficiency
   B. DOP HEPAP Method
   C. fractional efficiency test
   D. synthetic dust weight arrestance

_____ 42. The measure of an air filter's ability to remove man-made dust from test air describes _____.
   A. atmospheric dust spot efficiency
   B. DOP HEPAP Method
   C. fractional efficiency test
   D. synthetic dust weight arrestance

_____ 43. An advanced technique used to measure air filter performance in which extremely accurate equipment is used to count the number of particles trapped and also classify each based on size describes _____.
   A. atmospheric dust spot efficiency
   B. DOP HEPAP Method
   C. fractional efficiency test
   D. synthetic dust weight arrestance

_____ 44. The air filter efficiency test that uses small, white, sample filters to collect testing particles from the airstreams ahead of the test filter describes _____.
   A. atmospheric dust spot efficiency
   B. DOP HEPAP Method
   C. fractional efficiency test
   D. synthetic dust weight arrestance

Name _____

Match each of the parts of the electronic air cleaner with their functions.

_____ 45. Creates an electric field that ionizes particles

_____ 46. Blocks large particles

_____ 47. Attracts and holds positively charged particles

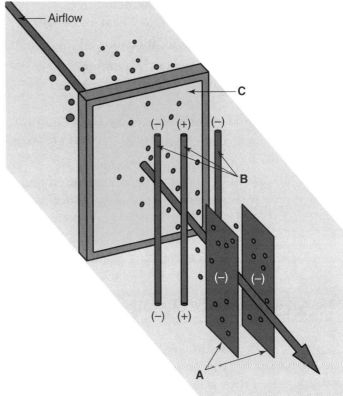

Airflow

C

B

A

9

_____ 48. *True or False?* Electronic air cleaners generally use a low voltage to charge particles in the air.

_____ 49. Before an electronic air cleaner's filter is removed, the power should be _____.

    A. connected
    B. decreased
    C. disconnected
    D. increased

_____ 50. Ionizing air purifiers may produce _____.

    A. carbon dioxide
    B. carbon monoxide
    C. nitrogen oxide
    D. ozone

_____ 51. Wavelengths within the _____ band of UV light are effective germicidal light wavelengths.

    A. A
    B. B
    C. C
    D. D

_____ 52. Review manufacturer instructions before installing ultraviolet lights, as they can be harmful to _____.

A. evaporator coils
B. eyes and skin
C. sheet metal ductwork
D. stainless steel heat exchanger

_____ 53. Common locations for UV light installations include the following, *except* _____.

A. above the evaporator coil
B. at each supply air register
C. in the return air duct
D. under or between the evaporator coil

_____ 54. To verify whether a UV light is operational, _____.

A. leave the light in place and check the lamp light indicator
B. override any safety shutoff and look at the illuminated light within the duct
C. override any safety shutoff and manually feel the light for warmth
D. pull the light out and connect power to visually confirm illumination

_____ 55. *True or False?* Ultraviolet lights used in cooling coils and pan areas can reduce the need for chemical cleaning agents.

28.5 Indoor Air Quality Systems

_____ 56. A natural exchange of air through leakage into a building is known as _____.

A. exfiltration
B. filtration
C. infiltration
D. ventilation

_____ 57. A natural passage of air leaving a building through doors, windows, and other construction joints is referred to as _____.

A. exfiltration
B. filtration
C. infiltration
D. ventilation

Name _____

> *Identify the parts of the home air quality system shown in the illustration. Match the letter of each part to its name.*

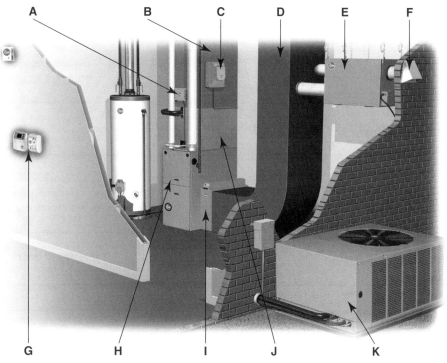

Rheem Manufacturing Company

_____ 58. Condensing unit

_____ 59. ERV/HRV

_____ 60. Filtration device

_____ 61. Fresh air intake

_____ 62. Furnace/air handler

_____ 63. Germicidal light

_____ 64. Humidifier

_____ 65. Indoor coil

_____ 66. Return air duct

_____ 67. Supply duct

_____ 68. Thermostat

_____ 69. To reduce the energy loss that would be required to recondition incoming fresh air, an HVAC system can route incoming fresh air through a(n) _____.

    A. boiler

    B. furnace

    C. grid of heating elements

    D. HRV or an ERV

Critical Thinking

70. List some areas in HVACR systems that act as breeding grounds for the microorganisms found in bioaerosols. How can these areas be safely maintained?

_____

_____

_____

_____

71. Explain how the amount of pollen in the atmosphere is measured.

_____

_____

_____

_____

72. List at least three ways carbon monoxide can enter or contaminate a living or working space.

_____

_____

_____

73. List at least three potential sources of radon in a home.

_____

_____

_____

74. Explain at least two methods of preventing radon from entering a building.

_____

_____

_____

_____

_____

75. Over the years, building construction has trended toward more airtight buildings. Explain how this has affected indoor air quality over the years.

_____

_____

_____

_____

Name _____

76. List at least five proactive actions a building's staff can take to alleviate or reduce indoor environmental issues.

_____

_____

_____

_____

_____

_____

_____

_____

77. List at least four categories of individuals with an increased susceptibility to indoor air pollutants.

_____

_____

_____

_____

78. List at least three possible causes of building-related illness (BRI).

_____

_____

_____

79. Explain the primary difference between building-related illness and sick-building syndrome.

_____

_____

_____

_____

80. How does the performance of HEPA filters over a long span of time (weeks, months) compare to the performance of ionic and electronic air cleaners over a long span of time?

_____

_____

_____

_____

9

81. Clean air filters and dirty air filters can behave differently within a forced-air distribution system. Compared to having a clean air filter, explain how having a dirty air filter can affect energy costs.

_____

_____

_____

_____

_____

_____

82. List at least three things a technician should check when installing an air filter.

_____

_____

_____

_____

_____

83. Explain the difference between HRVs and ERVs.

_____

_____

_____

_____

CHAPTER 29

Air Distribution

Name _____

Date _____ Class _____

> *Carefully study the chapter and then answer the following questions.*

29.1 Air Properties and Behavior

_____ 1. As relative humidity increases, each pound of air will _____.

    A. increase in density
    B. increase in temperature
    C. occupy a larger space
    D. occupy a lesser space

_____ 2. As the temperature rises for a pound of air at a set humidity level, its density _____.

    A. decreases
    B. increases
    C. remains the same
    D. None of the above.

_____ 3. *True or False?* The gases that compose air have a definite mass; therefore, they also have weight.

_____ 4. The heat or energy stored within moisture in the air is _____ heat.

    A. chemical
    B. latent
    C. optical
    D. sensible

_____ 5. *True or False?* Warm air has the potential to hold more moisture than cold air.

_____ 6. *True or False?* An evaporative cooling system decreases the rh level of the air while raising its db temperature.

_____ 7. *True or False?* An air distribution system prevents stratification by keeping the air moving.

29.2 Air Circulation

_____ 8. Temperature difference between return and supply air is greatest in _____ systems.
   A. dehumidifier
   B. heat pump
   C. standard air-conditioning
   D. standard heating

_____ 9. Air delivered to a room from a supply duct is called _____ air.
   A. exhaust
   B. primary
   C. return
   D. secondary

_____ 10. Throw is the distance that air entering a room travels from the duct outlet before it slows to _____ fpm.
   A. 1
   B. 50
   C. 100
   D. 125

_____ 11. A room will have _____ if there is more air leaving than entering.
   A. balanced pressure
   B. negative pressure
   C. perfect vacuum
   D. positive pressure

_____ 12. *True or False?* Ducts providing air to a room should be located at the farthest distance from ducts removing air from a room.

29.3 Basic Ventilation Requirements

_____ 13. Common automatic methods of bringing fresh air into an average home's conditioned space include the following, *except* a(n) _____.
   A. ERV
   B. fresh air inlet damper
   C. HRV
   D. open window

_____ 14. *True or False?* Negative pressure decreases the infiltration at windows and doors.

_____ 15. Sufficient makeup air is needed for buildings with fuel-burning appliances, because proper combustion requires enough _____.
   A. argon
   B. nitrogen
   C. oxygen
   D. ozone

Name _____

_____ 16. A system in cooling mode should provide at least _____ of air per person.

    A. 15 cfm

    B. 40 cfm

    C. 100 cfm

    D. 1,000 cfm

_____ 17. Vented attics use the following for ventilation, *except* _____.

    A. attic fans

    B. HRVs

    C. louvers

    D. vents

_____ 18. Hot air rising in a vented attic can escape through the following, *except* _____.

    A. gable vents

    B. ridge vents

    C. roof ventilators

    D. soffit vents

_____ 19. Having upper and lower vents in a vented attic takes advantage of _____.

    A. centrifugal force

    B. natural convection

    C. solar radiation

    D. the Venturi effect

_____ 20. Opening windows and confirming that outdoor temperature is lower than indoor temperature describes preparation for turning on a(n) _____.

    A. air curtain

    B. fresh air inlet fan

    C. gable fan

    D. whole house fan

9

29.4 Air Ducts

_____ 21. *True or False?* Concerning ductwork and the conditioned space, as pressure difference increases, airflow becomes faster.

Identify each of the following types of duct systems.

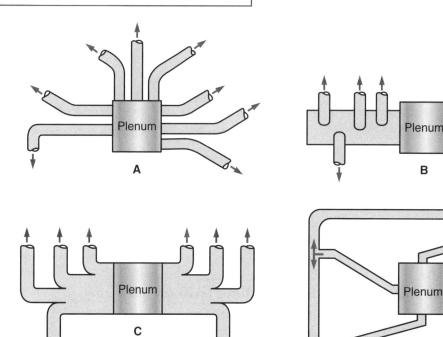

Goodheart-Willcox Publisher

_____ 22. Extended plenum system

_____ 23. Perimeter loop system

_____ 24. Radial duct system

_____ 25. Reducing trunk system

_____ 26. Ducts installed in unconditioned spaces, such as attics or crawl spaces, are usually _____ to reduce heat loss.
   A. galvanized
   B. insulated
   C. painted
   D. vented

_____ 27. The space between a false ceiling and the real ceiling is often used as a(n) _____.
   A. exhaust to outdoors
   B. inlet from outdoors
   C. return air plenum
   D. supply air plenum

_____ 28. To account for the expansion and contraction of a sheet metal duct, a(n) _____ is used.
   A. angle iron wrap
   B. fabric joint
   C. flexible rivet connector
   D. loosely made duct connection

Name _____

_____ 29. A round duct should be connected to a plenum using a(n) _____.

    A. fabric joint
    B. register boot
    C. snap lock
    D. starting collar

_____ 30. The crimped end of a round duct should always be facing the _____.

    A. air handler
    B. direction of airflow
    C. return air plenum
    D. supply air plenum

_____ 31. Ductboard is a panel that can be used to form ductwork. It is commonly made of _____.

    A. fiberglass
    B. plastic
    C. sheet metal
    D. sheetrock

_____ 32. The ratio of the duct's wide side to its narrow side is its _____.

    A. aspect ratio
    B. balance point
    C. cross-sectional ratio
    D. energy efficiency ratio

Match the terms with the descriptive phrase that applies to it. Answers will be used only once.

_____ 33. Bend in a duct that is usually 90°

_____ 34. Used in round ducts to control airflow

_____ 35. Uniformly spread the air leaving a duct

_____ 36. Nonadjustable fixture used to cover a return duct inlet

_____ 37. Primarily used in square or rectangular ducts to restrict or completely shut off airflow

_____ 38. Delivers air into a room using one or more adjustable airstreams

_____ 39. Used where an air path branches off in two directions

A. Butterfly damper

B. Diffuser

C. Elbow

D. Grille

E. Multiple-blade damper

F. Register

G. Split damper

9

_____ 40. *True or False?* Diffusers in air distribution systems spread the air leaving the duct more uniformly than registers do.

Identify the following types of dampers.

_____ 41. Butterfly damper

_____ 42. Multiple-blade damper

_____ 43. Split damper

Adjustable handle

A

Adjustable handle

B

Adjustable handle

C

Goodheart-Willcox Publisher

_____ 44. The following are ways to automatically position dampers, *except* for ____.

A. electric motors
B. manually
C. pneumatic motors
D. solenoids

_____ 45. An HVAC system that regulates the volume of air in its distribution system using dampers must use a ____ controller.

A. constant air volume (CAV)
B. variable air volume (VAV)
C. variable frequency drive (VFR)
D. variable refrigerant flow (VRF)

_____ 46. The type of fire damper that is required at locations in which fan pressure will be on during a fire is a(n) ____ fire damper.

A. fan-on
B. dynamic
C. static
D. variable flow

_____ 47. Automatic fire dampers should be installed in all ____ ducts in commercial and industrial buildings.

A. horizontal
B. return
C. supply
D. vertical

Name _____

_____ 48. The class of fire damper that can be used where a one-hour hold is required is a Class _____.
  A. A
  B. B
  C. C
  D. D

_____ 49. The class of fire damper that will hold back a fire indefinitely is a Class _____.
  A. A
  B. B
  C. C
  D. D

_____ 50. The class of fire damper that can be used when a two- to four-hour hold is required is a Class _____.
  A. A
  B. B
  C. C
  D. D

29.5 Duct Sizing

_____ 51. The pressure difference between the blower inlet and outlet is called the _____.
  A. aspect ratio (APR)
  B. available static pressure (ASP)
  C. device pressure loss (DPL)
  D. external static pressure (ESP)

_____ 52. The amount of friction or resistance used to design the longest duct run is called _____.
  A. aspect ratio (APR)
  B. available static pressure (ASP)
  C. device pressure loss (DPL)
  D. external static pressure (ESP)

_____ 53. Effective length of a duct run is calculated by adding the physical length of the ductwork and the effective length values of _____.
  A. the air filter and any other air-cleaning devices
  B. each fitting in the duct run
  C. the evaporator coil
  D. the blower

_____ 54. Friction loss is the _____ resulting from the air contacting the inside surface of the duct.
  A. Btu reduction
  B. pressure drop
  C. speed reduction
  D. temperature change

_____ 55. Friction loss is typically expressed in units of _____ per 100′ of duct.
  A. inches of mercury column
  B. inches of water column
  C. microns
  D. psi

_____ 56. *True or False?* Locating the air handler in a central location minimizes the lengths of the longest supply and return ducts.

_____ 57. *True or False?* When designing a duct system, ducts should always be installed in the end of a plenum.

_____ 58. Device pressure loss may result from the following, *except* a(n) _____.

    A. air filter
    B. compressor
    C. damper
    D. evaporator coil

_____ 59. The ACCA manual developed to be used to design both the supply and return ducts of residential buildings is _____.

    A. Manual D
    B. Manual J
    C. Manual N
    D. Manual S

29.6 Fans

_____ 60. An axial fan uses a _____ fan.

    A. centrifugal
    B. propeller
    C. Either A or B.
    D. Neither A nor B.

_____ 61. A radial fan uses a _____ fan.

    A. centrifugal
    B. propeller
    C. Either A or B.
    D. Neither A nor B.

_____ 62. The pressure generated on the exhaust side of a fan is _____.

    A. negative
    B. positive
    C. vacuum
    D. All of the above.

_____ 63. The pressure generated at the air inlet to a fan is _____.

    A. negative
    B. positive
    C. vacuum
    D. All of the above.

_____ 64. The air feed into a fan is called _____ draft.

    A. forced
    B. induced
    C. negative
    D. positive

Name _____

_____ 65. The exhaust from a fan is called _____ draft.

    A. forced

    B. induced

    C. negative

    D. positive

_____ 66. *True or False?* Radial flow fans may be directly driven by the fan motor or driven by a belt and pulley system.

29.7 Air Curtains

_____ 67. In commercial applications, an air curtain is installed to blow air _____.

    A. across a building's opening to outdoors

    B. across a cooling coil or evaporator

    C. through a furnace

    D. throughout a conditioned space

_____ 68. *True or False?* An air curtain is often installed with the intention of providing a barrier between two conditioned spaces maintained at different temperatures.

_____ 69. An air curtain has the purpose of reducing _____.

    A. accumulated stagnant air in a conditioned space

    B. heat gain and heat loss

    C. positive pressure in a building

    D. total pressure drop

Critical Thinking

9

70. Explain why humidifiers are used during cold weather. Also, explain what is happening to the air regarding moisture and heat when a humidifier is operating.

_____

_____

_____

_____

_____

71. Explain the consequences of installing a thermostat or humidistat too high on a wall. Also explain the consequences of installing a thermostat or humidistat too low on a wall. How will this affect occupants?

_____

_____

_____

_____

72. Explain the different ways that a blower can be controlled to change airflow in a non-zoned air distribution system.

_____

_____

_____

_____

73. Explain why it is best to bring replacement air into a building through an air system rather than bringing it in directly from outside.

_____

_____

_____

_____

_____

_____

74. A close friend lives in a hot and humid climate. Your friend's basement always seems to be humid. It tends to accumulate moisture, which puts the basement in danger of mold growth. What method or solution would you propose to solve this problem? Explain why you would recommend this solution, including any pros and cons.

_____

_____

_____

_____

_____

_____

75. Explain the different reasons why round ducts are overall more efficient than square or rectangular ducts.

_____

_____

_____

76. Explain the different causes of noise in air duct systems.

_____

_____

_____

_____

Name _____

77. Explain why large sections of sheet metal ducts should be cross-broken.

_____

_____

_____

78. Explain why a duct cross section is usually enlarged at a grille.

_____

_____

_____

79. List two ways of making airflow directional changes more gradual to minimize air turbulence within ductwork.

_____

_____

80. Explain how airflow from an axial flow fan differs from airflow from a radial flow fan.

_____

_____

_____

9

Notes

CHAPTER 30

Ventilation System Service

> *Carefully study the chapter and then answer the following questions.*

30.1 Airflow Measurement

_____ 1. The allowable draft rate in a room is _____ fpm.

    A. 15 to 25
    B. 50 to 100
    C. 250 to 350
    D. 400 to 600

_____ 2. Air stagnation occurs if air movement is less than _____ fpm.

    A. 15
    B. 25
    C. 50
    D. 250

_____ 3. The type of testing used to determine the airtightness of the building envelope is _____ testing.

    A. air handler flowmeter
    B. blower door
    C. duct blaster
    D. flow hood

_____ 4. *True or False?* When negative pressure is created indoors during testing, the higher-pressure outside air flows in through any cracks and openings.

_____ 5. An instrument used to direct and calculate all of the airflow through a duct at a given supply or return is a(n) _____.

    A. air handler flowmeter
    B. blower door
    C. duct blaster
    D. flow hood

_____ 6. An instrument that measures all of the airflow through a central HVAC system's air handler is a(n) _____.

    A. air handler flowmeter
    B. blower door
    C. duct blaster
    D. flow hood

_____ 7. An instrument used to create a pressure difference between indoors and outdoors is a(n) _____.

    A. air handler flowmeter
    B. blower door
    C. duct blaster
    D. flow hood

_____ 8. The initialism *WRT* is used to indicate the reference to which a pressure reading is being compared and stands for _____.

    A. which reference taken
    B. wind resistance tracing
    C. with respect to
    D. working real test

_____ 9. *True or False?* A tight building envelope has no gaps in its insulation or air barrier.

_____ 10. A blower door assembly uses a(n) _____ to measure pressure.

    A. anemometer
    B. gauge manifold
    C. manometer
    D. micron gauge

_____ 11. When the volume of air within a building is replaced with an equal volume of air, then a(n) _____ has occurred.

    A. air balancing
    B. air change
    C. air sweep
    D. duct blasting

_____ 12. The measured airflow through a blower door fan that is used to calculate air changes per hour is called _____.

    A. cfm_{25}
    B. cfm_{50}
    C. cfm_{60}
    D. cfm_{500}

_____ 13. The airflow value that is used when duct testing is _____.

    A. cfm_{25}
    B. cfm_{50}
    C. cfm_{60}
    D. cfm_{500}

_____ 14. *True or False?* A duct leak in a conditioned space is a greater concern than a duct leak in an unconditioned space.

30.2 Special Duct Problems and Duct Maintenance

_____ 15. The rigid structures that transmit vibrations are noise _____.

    A. amplifiers
    B. carriers
    C. reflectors
    D. sources

Name _____

_____ 16. What causes air turbulence that creates noise in a duct system?

    A. Elbows with built-in directional vanes
    B. High-velocity airflow
    C. Long-radius elbows
    D. None of the above.

_____ 17. A popping sound when a central HVAC system starts or stops is usually caused by _____.

    A. expansion or contraction of the duct as it warms up or cools
    B. a fan blade snapping off from the blower wheel
    C. a fuse blowing in the electrical box
    D. high-velocity airflow

_____ 18. Soft fabrics, such as drapes, curtains, and fabric-covered furniture, most often function as noise _____.

    A. absorbers
    B. amplifiers
    C. carriers
    D. sources

_____ 19. What may be the problem if, after removing a grille, a high-pitched sound continues?

    A. Air velocity is too low.
    B. The blower motor bracket is loose.
    C. Temperature change has deformed a section of ductwork.
    D. There is a sharp edge in the duct system.

_____ 20. *True or False?* Replacing a register with one that diffuses the air more narrowly can alleviate a draft from the register.

Match the letter for each step of cleaning ductwork in order from first to last.

_____ 21. Apply a microbial biocide spray to the inner walls of the ductwork.   A. Step 1

_____ 22. Remove loose particles using a vacuum.   B. Step 2

_____ 23. Rotate a duct sweeper to loosen dust, mold, and mildew.   C. Step 3

_____ 24. Balancing a system involves sizing ducts properly and adjusting _____ to ensure that each room receives the correct amount of air.

    A. the blower
    B. dampers
    C. room registers
    D. the thermostat

_____ 25. When a duct is too long, has too many elbows, or exhibits excess resistance to airflow for another reason, overcome this problem by installing a _____.

    A. blower door fan
    B. duct blaster
    C. duct booster
    D. flow hood

30.3 Fan Service

_____ 26. On an average residential HVAC system, dirt from fans should be removed every ____.

    A. six hours

    B. six days

    C. six weeks

    D. six months

_____ 27. Generally, HVAC fan bearings should be lubricated with ____ each year.

    A. one to two drops of oil

    B. 1/8 cup of oil

    C. 1/4 cup of grease

    D. 1 pint of oil

30.4 Filter Service

_____ 28. Replace an air filter if its pressure drop is more than ____ of the pressure drop across the system's blower.

    A. 5%

    B. 10%

    C. 25%

    D. 35%

_____ 29. Disposable filters should be checked and replaced every ____ unless special conditions require them to be replaced sooner.

    A. day

    B. week

    C. month

    D. three months

_____ 30. *True or False?* When washing an electronic air filter, an indication that the cleaner is removing contaminants is if the water used to wash the filter turns black.

_____ 31. To read the voltage on an electronic air cleaner, a(n) ____ must be used with a standard multimeter.

    A. 20 MΩ bypass resistor

    B. high-voltage probe

    C. in-line ammeter clamp

    D. thermocouple probe

Critical Thinking

32. What can smoke from smoke sticks, smoke guns, or smoke candles be used to check in a duct system?

_____

_____

_____

Name _____

33. Why should the outdoor intake end of a fresh air duct be fitted with a screen and bar combination?

_____

_____

_____

34. Why should smoke detectors be placed as far from high-velocity air ducts as possible?

_____

_____

35. List several areas where noise might originate in a duct system.

_____

_____

_____

_____

36. When it is not possible to reduce air velocity, explain the ways that duct noise can be reduced.

_____

_____

_____

37. When air velocity cannot be reduced, explain what can be done about supply air drafts.

_____

_____

_____

_____

38. List some of the benefits of a clean HVAC system.

_____

_____

_____

_____

39. Identify at least three conditions that indicate a need for HVAC system cleaning.

_____

_____

_____

_____

_____

9

40. What should be done when preparing a site for duct cleaning?

_____

_____

_____

_____

_____

41. List three types of tools that can be used for inspecting the inside of ductwork.

_____

_____

_____

42. What can occur to a building's conditioned spaces when its air distribution system is not balanced?

_____

_____

_____

_____

43. List at least three methods used to remove odors and vapors from a conditioned space.

_____

_____

_____

_____

_____

44. What problems can be caused by dirt buildup on fans and fan motors?

_____

_____

_____

_____

45. List several situations in which electronic air cleaners need servicing.

_____

_____

_____

CHAPTER 31

Ductless Air-Conditioning Systems

Name _____

Date _____   Class _____

Carefully study the chapter and then answer the following questions.

31.1 Principles of Cooling and Humidity Control

_____ 1. Comfort cooling systems use mechanical refrigeration to control temperature and ____.
   A. air cleanliness
   B. humidity
   C. odors
   D. None of the above.

_____ 2. The refrigerant evaporating temperature range of most comfort cooling systems is ____.
   A. 0°F to 15°F (–17.8°C to –9.5°C)
   B. 20°F to 30°F (–6.7°C to –1°C)
   C. 40°F to 50°F (4°C to 10°C)
   D. 65°F to 75°F (18°C to 24°C)

Match each of the components of a typical cooling unit with the correct term. Not all letters will be used.

_____ 3. Air return

_____ 4. Blower

_____ 5. Outdoor air

_____ 6. Exhaust air

_____ 7. Air filter

_____ 8. Evaporator

_____ 9. Damper

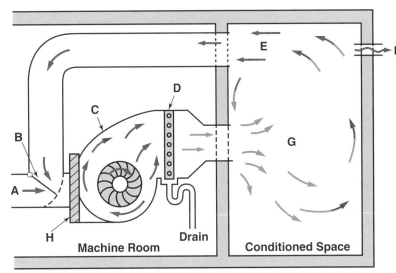

Goodheart-Willcox Publisher

_____ 10. Concerning air conditioning and psychrometrics, the lower the dry-bulb air temperature, _____.
   A. the higher its enthalpy
   B. the less moisture it can potentially hold
   C. the more moisture it can potentially hold
   D. All of the above.

_____ 11. What is the result when the dry-bulb temperature of set volume of air is reduced but its grains of moisture remain the same?
   A. Its enthalpy increases.
   B. Its relative humidity increases.
   C. Its wet-bulb temperature increases.
   D. All of the above.

The following psychrometric chart shows the conditions of air for a cooling system. This system is cooling conditioned air and also drawing in some fresh air from the hot outdoors. Both sets of air are being cooled for the conditioned space. Match each phrase with the correct lettered point or lettered line from the options based on the psychrometric chart.

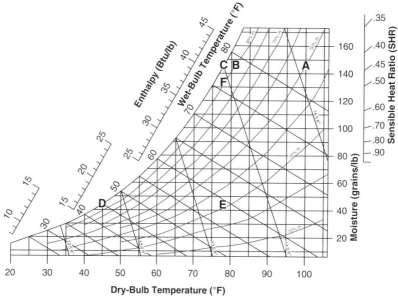

Goodheart-Willcox Publisher

_____ 12. Point representing the air in the conditioned space.

_____ 13. Line showing temperature rise from cooled air mixing with room air.

_____ 14. Point representing the mixture of outdoor air and conditioned air.

_____ 15. Line showing moisture being removed from the air.

_____ 16. Line showing mixed air being cooled to its saturation point.

_____ 17. Line showing the cooling of the outside air.

_____ 18. Line showing the rise in humidity from cooled air mixing with room air.

_____ 19. Point representing the condition of the outside air.

A. Line A to B

B. Line B to C

C. Line C to D

D. Line D to E

E. Line E to F

F. Point A

G. Point B

H. Point F

Name _____

31.2 Room Air Conditioners

_____ 20. Most window air conditioners have a(n) _____ as a refrigerant metering device.

    A. capillary tube

    B. EEV

    C. low-side float valve

    D. manual needle valve

_____ 21. As the moisture in a window air conditioner system evaporates, it helps to cool the _____.

    A. compressor and condenser

    B. compressor and fan motor

    C. evaporator and condenser

    D. evaporator and metering device

_____ 22. In a room air conditioner, the thermostat's sensing bulb is usually installed at the inlet of the _____.

    A. compressor

    B. condenser

    C. evaporator

    D. metering device

_____ 23. A standard room air conditioner is usually equipped with _____ to provide heating during cold weather.

    A. electric heating elements

    B. a gas burner

    C. an oil burner

    D. a wood burner

_____ 24. The first step in installing a window air conditioner is to _____.

    A. apply sealing compounds

    B. attach the fill boards

    C. install the sleeve

    D. position the air conditioner chassis in place

_____ 25. When installing a window air conditioner, the condensing side must _____.

    A. be level with the evaporating side

    B. tilt downward

    C. tilt upward

    D. rest on springs so it can bounce freely during operation

_____ 26. _True or False?_ When moving a room air conditioner into its sleeve, grip the coil and tubing of the evaporator.

_____ 27. _True or False?_ It is best to use a separate, dedicated electrical circuit for a room air conditioner.

10

Match each of the components of the window unit with the proper term.

_____ 28. Bottom channel

_____ 29. Leveling screw

_____ 30. Sash seal gasket

_____ 31. Side extension panel

_____ 32. Sill bracket

_____ 33. Top channel

_____ 34. Window seal gasket

Goodheart-Willcox Publisher

_____ 35. The main reason why metal blades shouldn't be used to clean condenser or evaporator coils is because _____.

    A. metal blades can poke through the coil tubing, causing leaks

    B. metal chips could form and later damage the fans

    C. the metal-to-metal contact can start corrosion through electrochemical action

    D. a spark might form and ignite leaking oil

_____ 36. PSC compressor motors have difficulty starting if the supplied voltage is low. To overcome this problem, a technician will often install a _____.

    A. buck-boost transformer

    B. motor starter

    C. relay and start capacitor

    D. starter coil

Name _____

Match each of the components of the window unit with the proper term.

_____ 37. Electrical control panel

_____ 38. Extension strap

_____ 39. Relay

_____ 40. Running capacitor

_____ 41. Starting capacitor

_____ 42. Tinnerman "U" nut

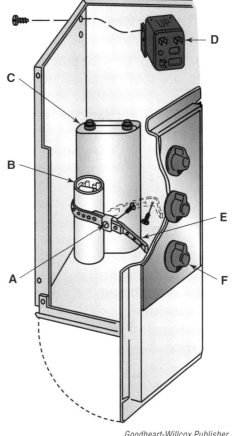

Goodheart-Willcox Publisher

_____ 43. How can a window unit be winterized?

    A. The air inlet and outlet grilles can be blocked with cardboard or flexible plastic sheeting.

    B. A storm sash can be custom built to fit around the air conditioner.

    C. A plywood box can be placed around the unit and held in place with caulking or rubber grommets.

    D. All of the above.

31.3 Packaged Terminal Air Conditioners

_____ 44. *True or False?* A packaged terminal air conditioner is installed with the evaporator side of the unit positioned outside the building.

_____ 45. The most common installation location for a PTAC is _____.

    A. in an exterior wall of a building

    B. in a central location within a building

    C. on the roof

    D. through a standard window opening

10

_____ 46. Electric resistance heat elements in a PTAC usually require _____ wiring.
   A. 5 Vdc
   B. 24 Vac
   C. 120 Vac
   D. 240 Vac

_____ 47. The method by which a PTHP changes between modes of operation (heating and cooling) is that _____.
   A. the compressor pumps in reverse
   B. the fans run in reverse
   C. a reversing valve changes the direction of refrigerant flow
   D. a series of dampers reverses airflow through the heat exchangers

_____ 48. A PTAC equipped with a gas-fired heater uses electronic ignition and either natural gas or _____ supplied to the system.
   A. diesel
   B. fuel oil
   C. kerosene
   D. LP gas

_____ 49. All types of PTACs use _____ compressors.
   A. belt-driven
   B. direct-drive
   C. hermetic
   D. None of the above.

_____ 50. PTACs are normally charged with _____.
   A. R-12 and R-123
   B. R-134a or R-410A
   C. R-500 and R-508
   D. R-717 and R-744

31.4 Console Air Conditioners

_____ 51. A primary difference between PTACs and console air conditioners is _____.
   A. their method of providing cooling
   B. the physical orientation of each system
   C. type of metering device
   D. type of compressor

_____ 52. *True or False?* Air-cooled console models must have air ducts to the outdoors for condenser cooling.

Name _____

Identify each of the components of the air-cooled console unit shown in the following illustration.

_____ 53. Blower

_____ 54. Compressor

_____ 55. Condenser

_____ 56. Condenser fan

_____ 57. Evaporator

_____ 58. Furnace section

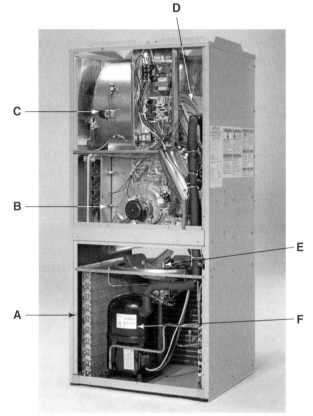

Suburban Manufacturing

_____ 59. *True or False?* Ducts can be connected to a console air conditioner to deliver conditioned air to areas with poor air circulation.

_____ 60. Console units typically use a(n) _____.

    A. AXV

    B. EEV

    C. high-side float

    D. TXV

31.5 Portable Air Conditioners

_____ 61. Warm condenser air from a portable air conditioner should not be vented ____.

    A. above a drop ceiling

    B. back into the cooled room

    C. into an adjacent room

    D. out of a window

_____ 62. Portable air conditioners are generally moved from one part of a large building to another part _____.

    A. by forklift

    B. by heavy moving equipment

    C. on a flatbed truck

    D. on their own wheels

_____ 63. *True or False?* Ceiling-mounted air conditioners have long runs of refrigerant-carrying tubes that need to be installed or brazed.

31.6 Multizone Ductless Split System

_____ 64. The indoor units of a multizone ductless split system are installed in separate rooms and are connected to the outdoor unit by a _____.

    A. line set

    B. mechanical linkage

    C. pneumatic tube

    D. radio frequency

_____ 65. Each of the indoor units of a multizone ductless split system has its own _____.

    A. blower

    B. evaporator

    C. metering device

    D. All of the above.

_____ 66. *True or False?* In a multizone ductless split heat pump system, the outdoor and indoor units can act as either condenser or evaporator, depending on the mode of operation.

_____ 67. *True or False?* In a building with multiple occupants, a major benefit of having multiple indoor units connected to a ductless system's outdoor unit is zoned temperature control.

_____ 68. A system that uses an EEV, a variable-speed compressor, and related controls can function as a _____ system.

    A. chiller

    B. heat pump

    C. variable air volume

    D. variable refrigerant flow

_____ 69. A multizone ductless system uses _____.

    A. an air filter in each of its indoor units

    B. an air filter in each of its outdoor units

    C. no air filters at all

    D. only one air filter

_____ 70. When an indoor unit's drain line is blocked, the overflow switch will _____.

    A. cycle on heating mode to evaporate the moisture

    B. open a solenoid valve in the secondary drain line

    C. run the indoor fan to evaporate the moisture

    D. shut off system operation

Name _____

_____ 71. The best way to prevent the growth of microorganisms is to place _____ in each indoor unit condensate pan.

    A. biocide tablets

    B. mint leaves

    C. plumber's grease

    D. WD-40

Critical Thinking

72. Explain the process of how a mechanical air-conditioning system removes moisture from the air to make the air more comfortable. What does the system do? How does this affect the air?

_____

_____

_____

_____

_____

_____

73. What is likely to happen if a window air conditioner is restarted immediately after cycling off? Explain why these things would happen.

_____

_____

_____

_____

_____

74. Describe the different ways and locations an outdoor unit for a multizone ductless split system may be installed.

10

_____

_____

_____

_____

Notes

CHAPTER 32

Residential Central Air-Conditioning Systems

Carefully study the chapter and then answer the following questions.

32.1 Central Air Conditioning

_____ 1. A split air-conditioning system is considered *split* because it splits its _____.

A. high-pressure refrigerant between two separate condensers
B. low-pressure refrigerant between two separate evaporators
C. refrigerant charge into low and high sides
D. system components into two separate locations

_____ 2. A split system's evaporator is most often found in the _____.

A. central air handler
B. condensing unit
C. return air duct
D. Any of the above.

_____ 3. A split system's compressor is most often found in the _____.

A. central air handler
B. condensing unit
C. return air duct
D. Any of the above.

_____ 4. A split system with a single air filter will usually have the air filter placed _____.

A. in the condensing unit
B. just after (downstream of) the blower
C. just after (downstream of) the evaporator
D. just before (upstream of) the blower

32.2 Split Systems

_____ 5. *True or False?* Some unitary systems come with components that are precharged with refrigerant.

10

_____ 6. Connections that can be made without losing refrigerant or getting air into a precharged system are _____.
   A. brazed swage joints
   B. flare joints
   C. quick-connect couplings
   D. threaded couplings

_____ 7. If an upflow gas furnace was installed in a residence and cooling was to be added, an evaporator would most likely be installed _____.
   A. in the plenum just above the furnace
   B. in the plenum just below the furnace
   C. just before (upstream of) the air filter
   D. None of the above.

_____ 8. A cooling system with an evaporator and condenser that are oversized for an application will result in a system that cycles off before it has provided adequate _____.
   A. cooling
   B. dehumidification
   C. humidification
   D. pressurization

Match each of the components of the air handler with the proper term.

_____ 9. Blower

_____ 10. Electronic controls

_____ 11. Furnace exhaust

_____ 12. Gas burners

_____ 13. Gas valve

_____ 14. Heat exchanger

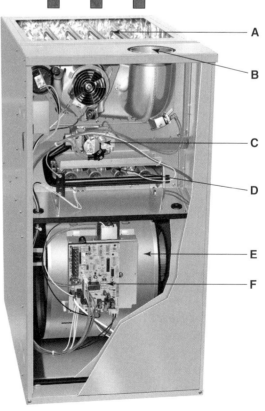

Tempstar

Name _____

> *Match each of the components with the proper term.*

_____ 15. Distributor

_____ 16. Liquid line connection (capped)

_____ 17. Sensing bulb

_____ 18. Suction line connection (capped)

_____ 19. Suction line manifold

_____ 20. TXV

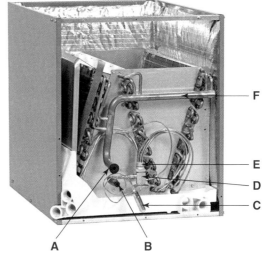

Rheem Manufacturing Company

32.3 Comfort Cooling Controls

_____ 21. The primary operational control of a building's split a/c system is the _____.

    A. circuit breaker

    B. internal motor overload

    C. thermal overloads

    D. thermostat

_____ 22. The primary device that turns on a compressor is the _____.

    A. contactor

    B. high-pressure switch

    C. low-pressure switch

    D. thermal overloads

10

Match each of the control components in this compressor circuit with the proper term.

_____ 23. 24 Vac

_____ 24. 240 Vac

_____ 25. Compressor

_____ 26. Contactor coil

_____ 27. Contactor contacts

_____ 28. High-pressure switch

_____ 29. Low-pressure switch

_____ 30. Overloads

_____ 31. Starting capacitor

_____ 32. Thermostat

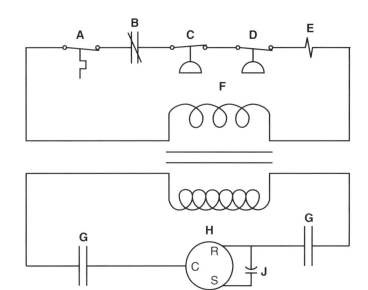

Goodheart-Willcox Publisher

32.4 Installing Central Air Conditioning

_____ 33. A condensing unit is usually installed about _____ away from a building.

    A. 3″ to 6″ (7.5 cm to 15 cm)
    B. 12″ to 24″ (31 cm to 61 cm)
    C. 60″ to 96″ (152 cm to 244 cm)
    D. 72″ to 108″ (183 cm to 274 cm)

_____ 34. A condensing unit should be installed in the following ways, *except* for _____.

    A. on a concrete slab outdoors
    B. on a manufactured pad outdoors
    C. in a specially vented room
    D. in an unvented attic

_____ 35. To prevent or discourage refrigerant theft, install _____ on a condensing unit's service valves.

    A. access valve locks
    B. blank-off plates
    C. condensate locks
    D. lockout relays

_____ 36. To reach an evaporator for cleaning, remove the _____ from the air handler.

    A. access valve plates
    B. blank-off plates
    C. cover plates
    D. lockout plates

Name _____

_____ 37. The sheet metal plates that support the evaporator coil and direct all plenum air through the evaporator coil are the _____.

    A. access valve plates

    B. blank-off plates

    C. cover plates

    D. lockout plates

_____ 38. A condensate line for a cooling system's evaporator should include a(n) _____ at the drain.

    A. air break

    B. P-trap

    C. steam trap

    D. U-trap

_____ 39. If a refrigerant line must be brazed to a service valve, be sure to _____.

    A. apply low-pressure flowing water through the inside of the connection

    B. direct a low-volume fan over the brazing area

    C. remove the Schrader valve core

    D. All of the above.

_____ 40. Which of the following statements about the installation of a condensing unit is *false*?

    A. The unit should be located in an inside corner of the exterior wall.

    B. The suction, liquid, and power lines should be as short as possible.

    C. If mounted outside, a concrete slab at least 4″ (10 cm) thick should be used.

    D. Outlet air from the unit should move in the same direction as the prevailing summer winds.

_____ 41. Which of the following statements regarding liquid line and capillary tube installation are *false*?

    A. A combination liquid line/capillary tube should be shortened if it seems long.

    B. The larger-bore tubing used for a combination liquid line/capillary tube reduces the chance of clogging from dirt or moisture.

    C. The liquid line may be used as a capillary tube.

    D. All of the above.

_____ 42. The best thing to do before servicing a condensing unit is to _____.

    A. disconnect the hot wire at the breaker box

    B. front seat each of the service valves

    C. open the electrical disconnect

    D. test the circuit with a clamp-on ammeter

32.5 Inspecting Central Air Conditioning Systems

_____ 43. Plants, shrubs, and branches should be cleared away from around an outdoor condensing unit primarily because _____.

    A. excess plants make a property look bad

    B. good clearance allows good airflow across the condenser coil

    C. good clearance makes service and maintenance work easier

    D. you can charge extra for that labor

10

_____   44. Outdoor condensing coils are best cleaned with _____.

    A. algaecide tablets
    B. generic cola
    C. phosphoric acid
    D. self-rinsing coil cleaner

_____   45. The most important wear point to check on contactors are the _____.

    A. contact points
    B. magnetic coil terminals
    C. mounting hardware
    D. power wire terminals

_____   46. The most important measurement to check and compare to manufacturer specifications on electric resistance heating elements is _____.

    A. current draw
    B. resistance
    C. temperature
    D. voltage drop

_____   47. What should the temperature differential be between the evaporator inlet and outlet?

    A. Between 0°F (0°C) and 7°F (4°C)
    B. Between 16°F (9°C) and 20°F (11°C)
    C. Between 20°F (11°C) and 35°F (19°C)
    D. Between 35°F (19°C) and 50°F (28°C)

32.6 Serving Central Air Conditioning Systems

_____   48. When preparing to steam clean an evaporator coil with clogged cooling fins, _____.

    A. remove the coil and plug all openings at once
    B. remove the coil and steam clean all openings before plugging
    C. steam clean the coil before removing it
    D. None of the above.

_____   49. During an annual cleaning of an outdoor condensing unit, bent fins should be _____.

    A. removed
    B. replaced
    C. straightened
    D. None of the above.

_____   50. A blockage in the condensate drain is most quickly and safely removed by _____.

    A. blowing out the drain using a blow torch
    B. blowing out the drain using compressed gas
    C. pouring caustic acid into the drain
    D. All of the above.

_____   51. When servicing a central HVAC system's air distribution network, _____.

    A. check for obstructions and sufficient airflow
    B. clean and sanitize ductwork as needed
    C. inspect all ductwork for leaks and broken dampers
    D. All of the above.

Name _____

32.7 Variable Refrigerant Flow (VRF) Systems

_____ 52. A fixed-speed compressor will _____.

A. always operate at maximum design rate
B. run at a fixed speed that is based on the temperature difference detected for each On cycle
C. vary its speed to match the load
D. All of the above.

_____ 53. VRF systems typically use a(n) _____ as the refrigerant metering device.

A. AEV
B. capillary tube
C. EEV
D. metering orifice

_____ 54. To properly manage the heat that needs to be picked up indoors and expelled outdoors, VRF systems typically use _____ indoor and outdoor fans.

A. automatic
B. single-speed
C. thermostatic
D. variable-speed

_____ 55. A disadvantage of VRF systems is that they cannot _____.

A. add humidity and control outdoor air changes
B. maintain individual temperature zones
C. regulate operation according to changing heat loads over time
D. All of the above.

_____ 56. The average ac motor's speed is primarily based on _____.

A. frequency and magnitude of the voltage
B. frequency of the voltage and number of poles
C. magnitude of the voltage and number of poles
D. None of the above.

_____ 57. The individual indoor units of a VRF system are connected _____.

A. back-to-front
B. in parallel
C. in reverse
D. in series

_____ 58. To avoid continuously cycling on and off, a VRF system will _____.

A. lower its operational capacity
B. lower or raise the setpoint by up to 10°F (5°C)
C. raise its operational capacity
D. shut off and illuminate its service warning light

10

Critical Thinking

59. List everything that a well-equipped central air-conditioning system is capable of doing to air.

_____

_____

_____

60. Explain the differences between a unitary system and a field-erected system. Explain why you might select one over the other.

_____

_____

_____

_____

61. What may happen if excessive pressure is applied to a wrench when tightening a quick-connect fitting?

_____

_____

62. List the four major steps for installing split central air-conditioning systems.

_____

_____

_____

_____

_____

63. List some building and local area changes that might require using a replacement air-conditioning system with a different capacity than the original.

_____

_____

_____

_____

_____

64. Explain some of the things that may happen if a suction line is _not_ insulated.

_____

_____

_____

_____

_____

Name _____

65. When should a central air-conditioning system be flushed?

_____

_____

_____

_____

66. Explain why it is important for a VRF system's compressor motor to operate above a minimum set speed. What could happen if its speed drops below the minimum set speed?

_____

_____

_____

_____

10

Notes

Commercial Air-Conditioning Systems

Carefully study the chapter and then answer the following questions.

33.1 Rooftop and Outdoor Units

_____ 1. Which of the following statements about rooftop units is false?

A. They are easy to install and do not take up valuable interior space.

B. They are factory-assembled and tested.

C. They are purchased as individual components and assembled on the roof of a building.

D. They may be heating only, cooling only, or both heating and cooling.

Match each of the components of the outdoor air-conditioning unit with the proper term.

_____ 2. Air filter

_____ 3. Blower

_____ 4. Burner

_____ 5. Compressor

_____ 6. Condenser

_____ 7. Evaporator

_____ 8. Flue

_____ 9. Humidifier

_____ 10. Metering device

_____ 11. Return air duct

_____ 12. Supply air duct

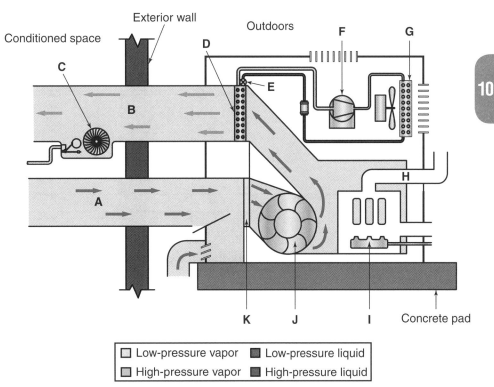

Goodheart-Willcox Publisher

10

_____ 13. When an RTU is calling for cooling, its air-side economizer will prevent the compressor from operating when _____.

    A. indoor air temperature is above the set point

    B. outdoor air is hot and humid

    C. outdoor air is cool and dry

    D. All of the above.

_____ 14. When an air-side economizer takes control of a system, the _____.

    A. compressor turns on

    B. condenser fan turns on

    C. fresh air inlet louvers open

    D. indoor blower turns off

_____ 15. An air-side economizer with enthalpy control will not turn on when _____.

    A. indoor air is hot

    B. indoor air is humid

    C. outdoor air is cool

    D. outdoor air is humid

Use the following psychrometric chart to answer questions 16 and 17.

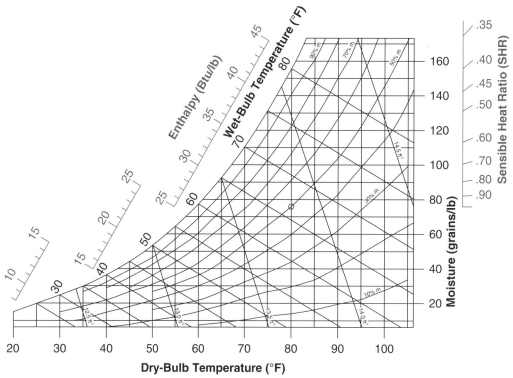

Goodheart-Willcox Publisher

_____ 16. Refer to the psychrometric chart. Which air sample has the highest heat content?

    A. 40°F (4.4°C) at 90% relative humidity.

    B. 50°F (10°C) at 70% relative humidity.

    C. 75°F (24°C) at 60% relative humidity.

    D. 90°F (32°C) at 10% relative humidity.

Name _____

_____ 17. Refer to the psychrometric chart. If an RTU's economizer is set for 50°F at 30% relative humidity, the air will have a heat content of approximately _____ Btu/lb or less.

    A. 9

    B. 15

    C. 20

    D. 28

_____ 18. In a commercial HVAC system in cooling mode, a thermostat's switch may directly control a _____.

    A. condenser fan

    B. liquid line solenoid valve

    C. low-pressure switch

    D. None of the above.

_____ 19. The purpose of motor limit controls is _____.

    A. adjusting equipment operation to meet demand

    B. general motor operation

    C. protection of the motor

    D. None of the above.

_____ 20. The purpose of sequential controls in commercial air conditioning is _____.

    A. adjusting equipment operation to meet demand

    B. general motor operation

    C. protection of the motor

    D. None of the above.

_____ 21. A commercial HVAC system that is controlled by pressurized air is a(n) _____ control system.

    A. economizer

    B. pneumatic

    C. VAV

    D. VRF

_____ 22. During installation of an RTU, be sure to properly install the _____ to prevent leaks into the building.

    A. acoustical material

    B. low-voltage control wiring

    C. resilient gasket

    D. All of the above.

_____ 23. An RTU with a gas furnace installation should include a gas line _____ outside the RTU's case.

    A. automatic vent

    B. check valve

    C. shutoff valve

    D. None of the above.

_____ 24. When using a portable ladder to work on a rooftop unit, _____.

    A. lean the ladder against the building at an angle so that the base is 1' away from the building for every 4' of vertical use

    B. extend the top of the ladder 3' above the edge of the roof

    C. use both hands on the ladder while climbing or descending

    D. All of the above.

10

_____ 25. To avoid excessive pressure drop, VRF systems should pump refrigerant vertically no more than _____ feet.

    A. 5

    B. 15

    C. 50

    D. 150

_____ 26. *True or False?* In new buildings that were specifically designed for use with a VRF system, air ducts can be smaller than normal, because the air only needs to be circulated and humidified.

33.2 Chillers

_____ 27. A chiller provides cooling using _____.

    A. chilled forced air

    B. a chilled water loop

    C. direct-expansion refrigerant coils

    D. None of the above.

_____ 28. Chiller systems are primarily used in _____.

    A. domestic appliances

    B. larger industrial and commercial applications

    C. motor vehicle cooling systems

    D. single-family residences

_____ 29. Modern compression chiller systems typically use _____ as a secondary refrigerant.

    A. ammonia

    B. oxygen

    C. R-12

    D. water, brine, or glycol solutions

_____ 30. When a chiller uses a water-cooled condenser, _____.

    A. the evaporator and condenser water loops share the same piping

    B. evaporator water and condenser water mix freely

    C. the two water circuits are kept separate and do not mix

    D. Both A and B.

_____ 31. Fans are used in cooling towers to _____.

    A. aid in evaporation

    B. chill the chiller water

    C. circulate the refrigerant

    D. All of the above.

_____ 32. Chiller efficiency is primarily based on _____.

    A. Btus absorbed per minute from all of the conditioned spaces combined

    B. compressor displacement rating

    C. gallons of water pumped per minute

    D. the temperatures of water entering and water leaving the chiller

Name _____

_____ 33. The primary purpose of a secondary refrigerant is to _____.

    A. absorb heat from another refrigerant
    B. absorb heat from a conditioned space
    C. cool a compressor
    D. expel heat into an unconditioned space

_____ 34. The average temperature difference of chiller water and its matching condenser's water should be about _____.

    A. 10°F (5.5°C)
    B. 20°F (11°C)
    C. 25°F (14°C)
    D. 35°F (19°C)

_____ 35. AHRI recommends that chillers operate at _____.

    A. 14°F (–10°C)
    B. 44°F (6.7°C)
    C. 64°F (17.8°C)
    D. 74°F (23°C)

_____ 36. Purge units are designed to remove noncondensables from a(n) _____.

    A. absorption chiller's secondary refrigerant circuit
    B. cooling tower's water circuit
    C. low-pressure chiller's primary refrigerant circuit
    D. All of the above.

_____ 37. *True or False?* If a well-maintained purge unit begins operating for longer periods of time than normal, a leak has probably developed.

_____ 38. A common time when a compressor's cylinders are unloaded is _____.

    A. at start-up
    B. during the Off cycle
    C. just before cycling off
    D. None of the above.

_____ 39. Cylinder unloading accomplishes _____.

    A. an increase of a compressor's capacity
    B. maximum refrigerant flow
    C. a reduction of a compressor's capacity
    D. None of the above.

_____ 40. One way of determining whether one or more cylinders of a compressor are unloaded without accessing the refrigerant circuit involves measuring _____.

    A. amperage
    B. resistance
    C. voltage
    D. None of the above.

10

_____ 41. Cylinder unloading is used on _____ compressors.

    A. centrifugal

    B. screw

    C. scroll

    D. reciprocating

_____ 42. To prevent pressure equalization during the Off cycle in a system with a scroll compressor, the best place to install a check valve would be _____.

    A. at the compressor's inlet

    B. at the compressor's outlet

    C. at the evaporator outlet

    D. in the liquid line

_____ 43. The capacity of screw compressors is controlled with _____.

    A. check valves

    B. cylinder unloading

    C. slide valves

    D. All of the above.

_____ 44. A centrifugal compressor creates a pressure difference in the refrigerant circuit using its _____.

    A. impeller

    B. intertwined scrolls

    C. matched helical rotors

    D. pistons, cylinders, and valves

_____ 45. The purpose of a compression chiller economizer is to improve system efficiency by _____.

    A. limiting the amount of refrigerant pulled into the compressor

    B. modulating the speed of compressor motor rotation

    C. separating oil from refrigerant to prevent it from interfering with heat transfer

    D. subcooling the high-side refrigerant

_____ 46. Lubrication for a centrifugal compressor is provided by _____.

    A. capillary action and wicking from a reservoir

    B. gravity through a drip system

    C. a pump driven by the compressor motor

    D. a separate motor used only for the oil pump

_____ 47. A centrifugal compressor's capacity control is provided by _____.

    A. cylinder unloading

    B. inlet guide vanes

    C. slide valves

    D. All of the above.

Name _____

Match each of the components of the centrifugal chiller system with the proper term.

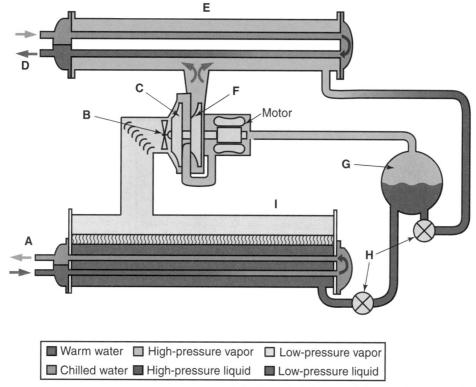

Carrier Corporation, Subsidiary of United Technologies Corporation

_____ 48. Chilled water

_____ 49. Chiller (evaporator)

_____ 50. Condenser water

_____ 51. Economizer

_____ 52. First-stage impeller

_____ 53. Inlet guide vanes

_____ 54. Metering device

_____ 55. Second-stage impeller

_____ 56. Water-cooled condenser

10

33.3 Cooling Towers

Match each of the components of the air-conditioning system with the proper term. Not all letters will be used.

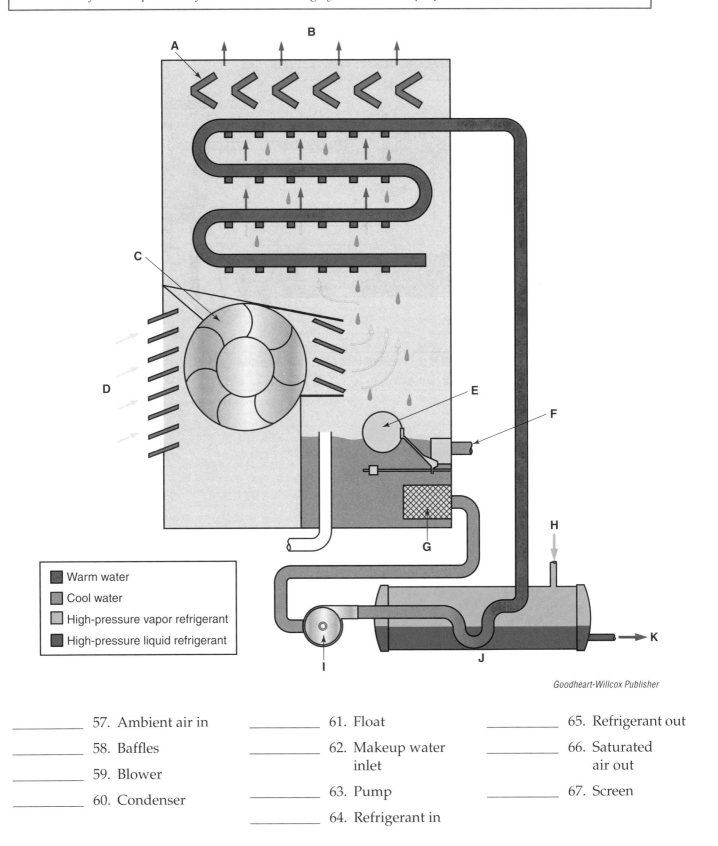

Warm water
Cool water
High-pressure vapor refrigerant
High-pressure liquid refrigerant

Goodheart-Willcox Publisher

_____ 57. Ambient air in

_____ 58. Baffles

_____ 59. Blower

_____ 60. Condenser

_____ 61. Float

_____ 62. Makeup water inlet

_____ 63. Pump

_____ 64. Refrigerant in

_____ 65. Refrigerant out

_____ 66. Saturated air out

_____ 67. Screen

Name _____

_____ 68. A cooling tower is a structure that removes heat directly from ____.

A. a building's potable water
B. chiller water
C. the conditioned space
D. an HVACR system's cooling water

_____ 69. The main method by which a cooling tower removes heat from water in warm weather is through ____.

A. cold storage underground
B. evaporation
C. mechanical refrigeration
D. an ice bath

_____ 70. As cooling tower water is lost, makeup water is added, which is controlled by the ____.

A. baffles
B. float
C. fill
D. water pump

_____ 71. As water travels downward through a cooling tower, its flow is generally guided by the ____.

A. baffles
B. float
C. fill
D. water pump

_____ 72. The purpose of a wet deck in a cooling tower is to ____.

A. increase contact between air and water
B. provide a work space for maintenance and service
C. serve as a reservoir for holding water
D. structurally support all the major parts of a cooling tower

_____ 73. A cooling tower that distributes water out of spray nozzles uses a ____.

A. centrifugal distributor
B. gravity flow basin
C. pressured water system
D. None of the above.

_____ 74. A cooling tower that distributes water passively out of precision holes uses a ____.

A. centrifugal distributor
B. gravity flow basin
C. pressured water system
D. None of the above.

_____ 75. *True or False?* Solenoid water valves are used when variable water flow control is necessary.

_____ 76. Pressure water valves have bellows that primarily react to ____ pressure.

A. ambient air (barometric)
B. high-side
C. low-side
D. water

10

_____ 77. *True or False?* A properly operating pressure water valve is able to maintain constant head pressure in a system.

_____ 78. If a low pressure pushing against the bellows of a pressure water valve began to increase in pressure, the valve should _____.

    A. move toward the closed position
    B. move toward the full open position
    C. not react at all
    D. None of the above.

_____ 79. A thermostatic water valve reacts to heat in _____.

    A. condenser inlet water
    B. condenser outlet water
    C. the liquid line
    D. the suction line

Critical Thinking

80. Many commercial air-conditioning systems have air-side economizers. Is it better to have a system with enthalpy control or without enthalpy control? Explain why.

_____

_____

_____

_____

81. Should a system's manual reset controls be replaced with automatic reset controls? Why or why not?

_____

_____

_____

_____

_____

_____

82. Explain what purpose a secondary emission-collection device serves on a chiller system. What does it prevent?

_____

_____

83. Briefly explain how capacity is maintained in multiple-compressor chiller systems. What is done when more capacity is needed? What is done when less capacity is needed?

_____

_____

_____

Name _____

84. Explain how a centrifugal compressor is able to reach such fast turning speed. What parts are used to do this? How are they used?

_____

_____

_____

_____

_____

_____

85. Weather conditions often affect the performance and efficiency of HVACR systems. Cooling towers are no exception to this. Is cooling tower operation more efficient when outdoor temperature is hot or when it is cold? Is it more efficient when outdoor air is dry or when it is humid? Explain why a cooling tower is more efficient in the weather conditions you have chosen.

_____

_____

_____

_____

_____

_____

86. Describe three ways in which a cooling tower loses water during operation.

_____

_____

_____

_____

_____

87. List at least three methods of cooling tower capacity control and explain how they work.

_____

_____

_____

_____

_____

_____

Notes

CHAPTER 34

Absorption and Evaporative Cooling Systems

Name _____

Date _____ Class _____

> *Carefully study the chapter and then answer the following questions.*

34.1 Absorption Refrigeration Systems

_____ 1. Absorption refrigeration systems use _____ energy to achieve refrigeration.

    A. heat
    B. magnetism
    C. mechanical
    D. vacuum

_____ 2. The function of an absorbant is to _____.

    A. fill any leaks in a sealed system
    B. get soaked up or absorbed by a refrigerant
    C. lubricate a machine's mechanical parts
    D. soak up or absorb a refrigerant

_____ 3. The purpose of an absorbant is to _____.

    A. lower pressure
    B. raise pressure
    C. raise temperature
    D. All of the above.

_____ 4. When a substance has an affinity for another substance, this means the first substance tends to _____ the other substance.

    A. absorb
    B. combust when in contact with
    C. dissociate with or expel
    D. fail to mix with

_____ 5. *True or False?* A strong solution contains very little refrigerant.

_____ 6. When a solid attracts and holds another substance on its outer surfaces, the process is called _____.

    A. absorption
    B. adsorption
    C. affinition
    D. anhydration

10

_____ 7. When two substances combine to form a uniform solution, the process is called _____.

    A. absorption

    B. adsorption

    C. affinition

    D. anhydration

_____ 8. An absorption system's component that raises refrigerant pressure is the _____.

    A. absorber

    B. condenser

    C. generator

    D. pump

_____ 9. An absorption system's component that circulates liquid refrigerant from the low side to the high side is the _____.

    A. absorber

    B. condenser

    C. generator

    D. pump

_____ 10. An absorption system that shuts off once its charge of fuel is expended is a(n) _____ absorption system.

    A. continuous-cycle

    B. cyclical

    C. intermittent

    D. sporadic

_____ 11. An absorption system that operates by the application of a limited amount of heat is a(n) _____ absorption system.

    A. continuous-cycle

    B. cyclical

    C. intermittent

    D. sporadic

34.2 Absorption Cooling Systems

_____ 12. When a residential absorption chiller begins operation, it starts by heating the _____.

    A. chilled water circuit

    B. evaporator

    C. mixed solution

    D. refrigerant heat exchanger

_____ 13. In a residential absorption chiller, the first pressure drop of ammonia is when it passes through a restrictor while entering the _____.

    A. absorber

    B. condenser

    C. evaporator

    D. heat exchanger

Name _____

_____ 14. In a residential absorption chiller, the weak solution undergoes a pressure drop when it passes through a restrictor while entering the _____.

    A. absorber
    B. condenser
    C. evaporator
    D. heat exchanger

_____ 15. In a residential absorption chiller, the strong solution returns to the generator by _____.

    A. centrifugal force
    B. the force of the solution pump
    C. gravity from the weight of condensation
    D. high pressure from high heat

_____ 16. To avoid the problems of a corrosive reaction in residential absorption systems, _____ should only be used to circulate warmed or chilled water and nothing else.

    A. aluminum
    B. copper and copper alloys
    C. stainless steel
    D. steel

_____ 17. The main component that allows an absorption system to function as a heat pump is its _____.

    A. change-over valve
    B. inverter
    C. reversing valve
    D. two-way pump

> *Match each of the absorption systems with the statement that best describes it.*

_____ 18. An absorption system that is directly heated by a burner or other heating element.

_____ 19. A system that uses a secondary, smaller, low-temperature generator to recover some of the waste heat from the primary generator.

_____ 20. System that does not have a secondary generator.

_____ 21. A system in which the heat that powers the system is produced somewhere outside the system.

A. Direct-fired absorption system

B. Double-effect absorption system

C. Indirect-fired absorption system

D. Single-effect absorption system

10

Match each of the components of the residential absorption chiller system with the proper term.

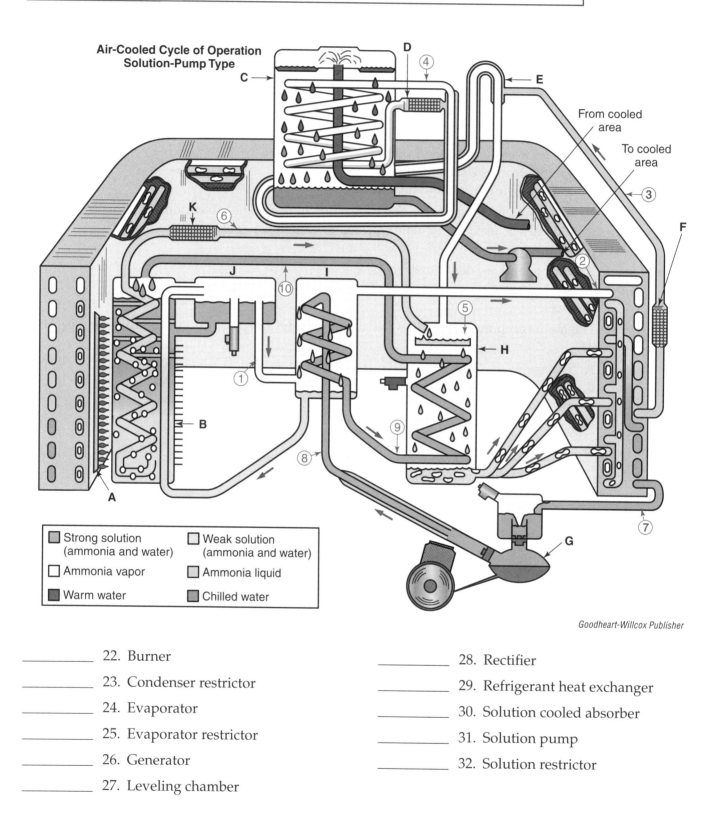

Air-Cooled Cycle of Operation Solution-Pump Type

Legend:
- Strong solution (ammonia and water)
- Weak solution (ammonia and water)
- Ammonia vapor
- Ammonia liquid
- Warm water
- Chilled water

From cooled area
To cooled area

Goodheart-Willcox Publisher

_____ 22. Burner

_____ 23. Condenser restrictor

_____ 24. Evaporator

_____ 25. Evaporator restrictor

_____ 26. Generator

_____ 27. Leveling chamber

_____ 28. Rectifier

_____ 29. Refrigerant heat exchanger

_____ 30. Solution cooled absorber

_____ 31. Solution pump

_____ 32. Solution restrictor

Name _____

_____ 33. *True or False?* A decrease in the temperature of the chilled water leaving a residential absorption chiller will decrease the overall capacity of the system.

_____ 34. *True or False?* A decrease in the temperature of ambient air decreases the overall capacity of a residential absorption chiller that is air cooled.

Match each of the components of the absorption heat pump (in heating mode) with the proper term.

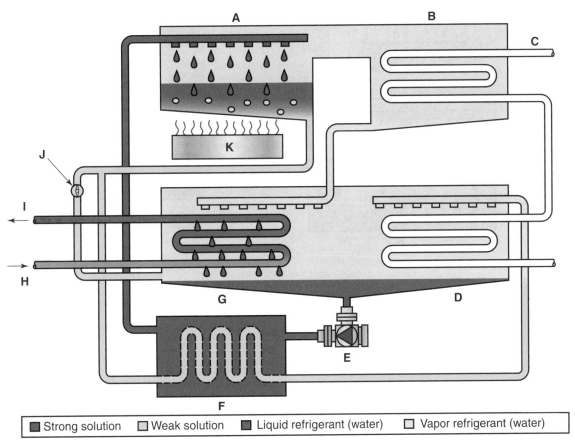

| ■ Strong solution | □ Weak solution | ■ Liquid refrigerant (water) | □ Vapor refrigerant (water) |

Goodheart-Willcox Publisher

_____ 35. Absorber

_____ 36. Change-over valve (open)

_____ 37. Condenser

_____ 38. Cooling water (off)

_____ 39. Evaporator

_____ 40. Generator

_____ 41. Heat exchanger

_____ 42. Heat source

_____ 43. Heated water in

_____ 44. Heated water out

_____ 45. Pump

10

Match each of the components of the direct-fired absorption chiller with the proper term.

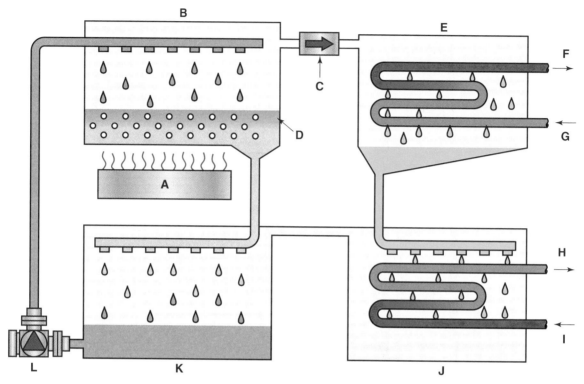

Goodheart-Willcox Publisher

_____ 46. Absorber

_____ 47. Burner

_____ 48. Check valve

_____ 49. Chilled water in

_____ 50. Chilled water out

_____ 51. Condenser

_____ 52. Cooling water in

_____ 53. Cooling water out

_____ 54. Evaporator

_____ 55. Generator

_____ 56. Pump

_____ 57. Refrigerant boiled out of solution

Name _____

> *Match the components of the ammonia absorption system (in heating cycle) to the proper terms.*

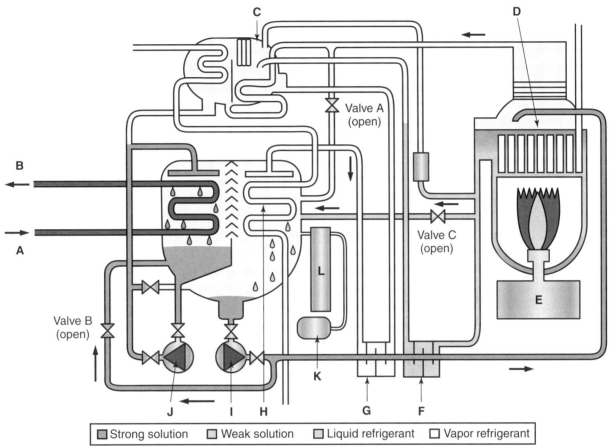

| ■ Strong solution | ■ Weak solution | ■ Liquid refrigerant | □ Vapor refrigerant |

Goodheart-Willcox Publisher

_____ 58. Absorbant pump

_____ 59. Absorber

_____ 60. Burner

_____ 61. High-temperature generator

_____ 62. High-temperature heat exchanger

_____ 63. Hot water inlet

_____ 64. Hot water outlet

_____ 65. Low-temperature generator

_____ 66. Low-temperature heat exchanger

_____ 67. Purge pump

_____ 68. Purge unit

_____ 69. Refrigerant pump

_____ 70. The upper shell of a lithium bromide absorption chiller contains the _____.

    A. absorber and condenser

    B. absorber and evaporator

    C. evaporator and condenser

    D. generator and condenser

_____ 71. The lower shell of a lithium bromide absorption chiller contains the _____.
A. absorber and condenser
B. absorber and evaporator
C. evaporator and condenser
D. generator and condenser

_____ 72. *True or False?* If the correct operating conditions are maintained inside a lithium bromide chiller, the lithium bromide always stays in vapor form.

_____ 73. *True or False?* Some lithium bromide absorption systems rely only on gravity to circulate the liquids through the system.

_____ 74. The use of energy by-products from one process as the primary energy source for another process is called _____.
A. absorption
B. cogeneration
C. regeneration
D. All of the above.

_____ 75. A combined heat and power (CHP) plant produces _____.
A. cold water and steam
B. electricity and mechanical force
C. electricity and useful steam
D. mechanical force and steam

34.3 Absorption System Service

_____ 76. An ammonia-based absorption chiller is usually charged with the following substances, *except* _____.
A. ammonia
B. corrosion inhibitor
C. distilled water (pH 6.0+)
D. lithium bromide

_____ 77. Ingredients used in a typical charge for use in a lithium bromide-based system include the following substances, *except* _____.
A. corrosion inhibitor
B. lithium bromide solution
C. octyl alcohol
D. sulfur dioxide

_____ 78. Leak testing a lithium bromide system generally involves _____.
A. bubble solution after charging with nitrogen or helium
B. injecting a fluorescent dye and visual detection
C. performing a temperature survey across all fluid lines
D. using an electronic leak tester along all fluid lines

_____ 79. Proper service techniques are important, because once air enters a lithium bromide system, the solution becomes _____.
A. highly corrosive
B. less able to absorb and mix
C. worse at transferring heat
D. None of the above.

Name _____

34.4 Evaporative Cooling

_____ 80. In evaporative cooling, heat is displaced into _____.
   A. air exhausted from the conditioned space
   B. a body of water
   C. evaporating refrigerant
   D. evaporating water

_____ 81. Evaporative cooling systems are most effective in climates that are _____.
   A. hot and dry
   B. hot and humid
   C. Both of the above.
   D. None of the above.

_____ 82. An arrangement of distribution water pipes installed above a tight lattice of fibers that covers an opening at one end of a building is generally used to cool a_____.
   A. classroom
   B. foundry
   C. greenhouse
   D. restaurant

_____ 83. Wet roof cooling is used primarily in climates where ambient temperature is _____.
   A. high and relative humidity is low
   B. low and relative humidity is high
   C. moderate and relative humidity is high
   D. None of the above.

_____ 84. Roof pond cooling maintains a body of water that stands _____ on the surface of a roof.
   A. 2 cm to 3 cm (20 mm to 30 mm)
   B. 2″ to 3″ (50 mm to 75 mm)
   C. 2′ to 3′ (61 cm to 91 cm)
   D. Any of the above.

_____ 85. Roof mist cooling systems have the potential to reduce the flow of solar heat from the roof into the building's interior by more than _____%.
   A. 5
   B. 25
   C. 50
   D. 90

_____ 86. Water flow on a roof mist cooling system is controlled by _____.
   A. float valves
   B. solenoid valves
   C. thermostatic expansion valves
   D. All of the above.

_____ 87. Roof mist cooling system operation is generally based on _____.
   A. conductivity switches
   B. float sensors
   C. humidity sensors
   D. temperature sensors

10

Critical Thinking

88. List five heat sources that can be used by absorption cooling systems. If your home had an absorption air conditioner, what heat source would you prefer? Explain why.

_____

_____

_____

_____

89. Are bubble leak tests used with ammonia absorption systems? Explain why or why not.

_____

_____

_____

90. Briefly explain the difference between single-effect absorption systems and double-effect absorption systems.

_____

_____

_____

_____

_____

91. In your own words, explain how a wet roof cooling system cools a structure.

_____

_____

_____

_____

_____

_____

CHAPTER 35

Humidity Control

Name _____

Date _____ Class _____

> *Carefully study the chapter and then answer the following questions.*

35.1 Humidity Levels and Comfort

_____ 1. A ratio of the moisture content in the air compared to the maximum amount of moisture that the air can hold at its current temperature describes _____.

    A. enthalpy
    B. dew point
    C. relative humidity
    D. sensible heat ratio (SHR)

_____ 2. A sample of air is able to hold more moisture _____.

    A. the higher its dry-bulb temperature
    B. the lesser its volume
    C. the lower its dry-bulb temperature
    D. None of the above.

_____ 3. Human comfort requires a relative humidity between _____.

    A. 10% and 20%
    B. 30% and 50%
    C. 75% to 90%
    D. 90% and 100%

_____ 4. *True or False?* Air changes per hour (ACH) refers to the number of times the entire volume of air in a conditioned space is replaced in one hour.

_____ 5. A humidistat is a control device that responds to _____.

    A. airflow
    B. air pressure
    C. humidity
    D. temperature

_____ 6. A humidifier's solenoid valve opens when the _____ running.

    A. compressor is
    B. dehumidifier is
    C. furnace and blower are
    D. HRV/ERV is

10

_____ 7. An electromechanical humidistat's hygroscopic element responds to a decrease in humidity by ____.
    A. bending in an arc
    B. shrinking
    C. straightening out
    D. stretching

35.2 Types of Humidifiers

Match each of the humidifiers with the statement that best describes it.

_____ 8. A type of humidifier that uses its own heat source to create steam.

_____ 9. A type of central humidifier that is installed through the bottom of ductwork.

_____ 10. A category of humidifiers that project small water droplets into the air either by mechanically flinging water or by forcing water through a nozzle.

_____ 11. A unit with a series of spinning plastic disks with shallow grooves.

_____ 12. A category of humidifiers that add moisture to the air through evaporation.

_____ 13. A unit that forces water through an orifice to create a very fine mist.

_____ 14. A unit with a spinning disk that flings water against a diffuser, which atomizes the water on impact.

_____ 15. A unit with a water pan and rotating drum covered with an absorbent sleeve.

_____ 16. A type of central humidifier that is installed between the supply and return plenums.

A. Atomizing humidifier
B. Bypass humidifier
C. Evaporative humidifier
D. Impeller humidifier
E. Nozzle-type humidifier
F. Rotating disk humidifier
G. Rotating drum humidifier
H. Under-duct humidifier
I. Vaporizing humidifier

_____ 17. Which one of the four types of evaporative humidifiers can be used only in low-capacity applications?
    A. Fixed filter
    B. Plate
    C. Rotating disk
    D. Rotating drum

Name _____

_____ 18. Which humidifier uses a pad consisting of numerous layers of a tight metallic mesh wetted by water delivered from above?
A. Fixed filter
B. Piezoelectric
C. Plate
D. Vaporizing

_____ 19. Which humidifier uses a series of porous materials that wick water upward to be evaporated by passing air?
A. Fixed filter
B. Piezoelectric
C. Plate
D. Vaporizing

_____ 20. A seasonal damper on a bypass humidifier should be closed _____.
A. always
B. never
C. when weather is dry
D. when weather is humid

_____
Match each of the components of the rotating drum humidifier with the proper term.
_____

_____ 21. Drum

_____ 22. Drum motor

_____ 23. Float

_____ 24. Humidistat

_____ 25. Overflow drain

_____ 26. Power line

_____ 27. Solenoid valve

_____ 28. Warm air duct

_____ 29. Water line

_____ 30. Water pan

_____ 31. Water vapor

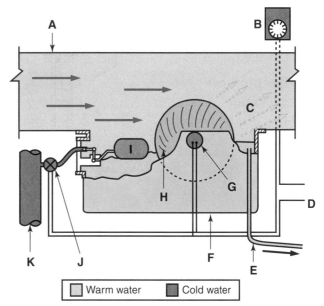

Warm water   Cold water

Goodheart-Willcox Publisher

10

> *Match each of the components of the under-duct rotating drum humidifier with the proper term.*

_____ 32. Absorbent filter

_____ 33. Drain connection

_____ 34. Drum motor

_____ 35. Water level adjustment screw

_____ 36. Water pan

_____ 37. Water supply connection

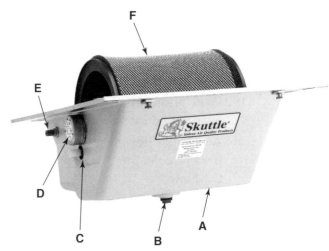

Skuttle IAQ Products

> *Match each of the components of the piezoelectric humidifier.*

_____ 38. Air inlet

_____ 39. Air outlet

_____ 40. Blower

_____ 41. Crystal plate

_____ 42. Electric wave generator

_____ 43. Pump

_____ 44. Water pan

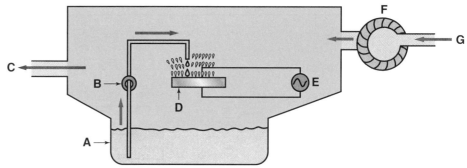

Goodheart-Willcox Publisher

> *Match each of the definitions with the letter for that type of water.*

_____ 45. Water that has been treated to remove minerals.

_____ 46. Natural, untreated water (such as rainwater) with a low mineral content.

_____ 47. Untreated water with 5 to 15 grains/gal. of mineral content.

_____ 48. Water in which an ion exchange has replaced unwanted minerals with water-soluble sodium salts.

_____ 49. Untreated water with over 15 grains/gal. of mineral content.

A. Demineralized water

B. Medium hard water

C. Soft water

D. Softened water

E. Very hard water

Name _____

_____ 50. The ideal water to use in humidifiers to prevent both duct deposits and foreign matter from entering the airstream is _____.

    A. city water

    B. distilled water

    C. softened water

    D. well water

_____ 51. To humidify a single room (rather than an entire building), it is best to use a(n) _____.

    A. electric heater

    B. hydronic heating system

    C. portable humidifier

    D. steam heating system

35.3 Dehumidifying Equipment

Match each of the terms with the components of the room dehumidifier.

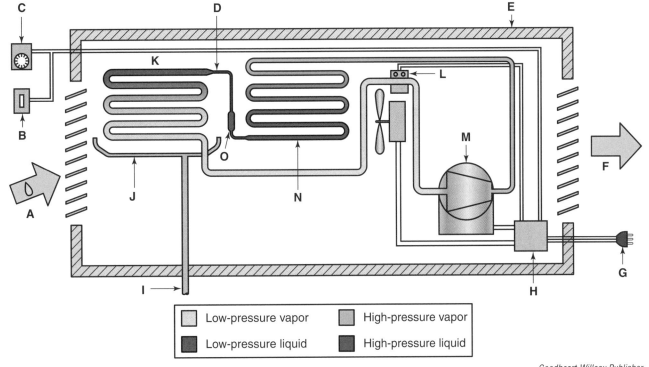

Goodheart-Willcox Publisher

_____ 52. Capillary tube

_____ 53. Casing

_____ 54. Compressor

_____ 55. Condensate pan

_____ 56. Condenser

_____ 57. Drain

_____ 58. Dry air out

_____ 59. Evaporator

_____ 60. Filter-drier

_____ 61. Frost control

_____ 62. Humid air in

_____ 63. Humidistat

_____ 64. Motor control

_____ 65. On-off switch

_____ 66. Power plug

_____ 67. Dehumidifiers remove moisture from the air by maintaining a coil _____ of the air.

    A. above the dew point
    B. above the vaporization point
    C. below the dew point
    D. below the vaporization point

_____ 68. A float switch on a dehumidifier is intended to _____.

    A. refill the water reservoir when it is low
    B. start operation of the dehumidifier
    C. start the blower
    D. stop operation of the dehumidifier

_____ 69. Dehumidification may be accomplished with special chemicals called _____.

    A. desiccants
    B. humidicants
    C. hygroscopes
    D. piezoelectric crystals

_____ 70. Special chemicals remove moisture from the air by _____.

    A. adsorption
    B. circulation
    C. evaporation
    D. high heat

35.4 Servicing and Installing Humidifiers

_____ 71. A humidifier must be mounted level if the humidifier has a _____.

    A. motor
    B. nozzle
    C. seasonal damper
    D. water reservoir

_____ 72. A humidifier's water feed line should be tapped into a building's _____.

    A. cold water line
    B. drain waste vent pipe
    C. furnace vent pipe
    D. gas line

_____ 73. The drainage outlet for a humidifier should be at least _____ the open drain.

    A. 1″ above
    B. 6″ above
    C. 1′ above
    D. 6′ above

_____ 74. _True or False?_ Water treatment compounds can be used regularly to prevent bacterial growth, algae, or scale buildup in a humidifier.

Name _____

Critical Thinking

75. Explain how high humidity affects hygroscopic materials. Describe evidence in an indoor environment that indicates high humidity.

_____

_____

_____

_____

_____

76. Explain some physical symptoms in humans in areas of low relative humidity.

_____

_____

_____

_____

77. Explain how low humidity affects hygroscopic materials. Explain what can happen to such materials (like wood) in low relative humidity.

_____

_____

_____

_____

_____

78. Explain the basic operation of a rotating drum humidifier.

_____

_____

_____

_____

_____

79. List five common sources of humidity. What can be done to mitigate the moisture contributed to the air by these sources?

_____

_____

_____

_____

_____

_____

80. Explain why the temperature of air entering a dehumidifier and air leaving a dehumidifier is so close in value. Why aren't the inlet and outlet air temperatures significantly different?

_____

_____

_____

_____

_____

_____

81. Describe some things that might cause a dehumidifier's float switch to be actuated. Should any service or maintenance be performed for these causes? Explain why or why not.

_____

_____

_____

_____

_____

_____

CHAPTER 36

Name _____

Date _____ Class _____

Carefully study the chapter and then answer the following questions.

36.1 What Is a Thermostat?

_____ 1. In simplest terms, a thermostat is a sensing device that reacts to _____.

A. airflow
B. humidity
C. pressure
D. temperature

_____ 2. A thermostat that remembers set points and adapts system operation and cycles accordingly is a _____ thermostat.

A. learning
B. master
C. power-stealing
D. Wi-Fi

_____ 3. The type of HVAC system operation that regulates the increase or decrease of intensity and can result in significant energy usage reduction over time is _____ operation.

A. lockout
B. modulating
C. on-off
D. None of the above.

_____ 4. The longer a bimetal coil is, the _____.

A. less bending movement it can produce
B. longer it will stretch
C. more bending movement it can produce
D. shorter it will stretch

_____ 5. A switch used with a bimetal coil that consists of a pair of contacts in a glass tube that changes position based on magnets is a _____ switch.

A. mercury
B. reed
C. rod and tube
D. thermocouple

10

_____ 6. A switch used with a bimetal coil that consists of a glass tube containing wire contacts and a puddle of liquid metal and that changes position based on its physical position is a _____ switch.

    A. mercury
    B. reed
    C. rod and tube
    D. thermocouple

_____ 7. Electronic thermostats use a _____ to sense air temperature.

    A. bimetal helix
    B. sensing bulb and bellows
    C. sensing bulb and diaphragm
    D. thermistor

_____ 8. In the event of a mercury spill, _____.

    A. call local HAZMAT authority
    B. flush the mercury down the toilet
    C. scoop up the mercury by hand
    D. vacuum the mercury puddle

Match the letter for each of the sensing elements with its name.

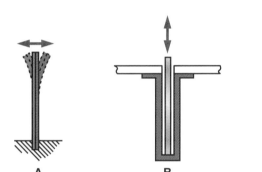  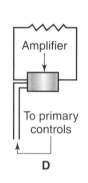  

    A   B   C   D   E   F

Goodheart-Willcox Publisher

_____ 9. Bimetal coil

_____ 10. Bimetal helix

_____ 11. Bimetal strip

_____ 12. Sensing bulb and bellows

_____ 13. Sensing bulb and diaphragm

_____ 14. Thermistor

Name _____

36.2 Types of Thermostats

_____ 15. The difference between a heating-only thermostat with two wires and a heating-only thermostat with three wires is the addition of a(n) _____.

    A. automatic venting circuit
    B. cut-out circuit
    C. holding circuit
    D. reversing circuit

_____ 16. For a heating thermostat, system overshoot occurs when temperature _____.

    A. drops below the cut-in value
    B. fails to reach the cut-in value
    C. fails to reach the cut-out value
    D. rises above the cut-out value

_____ 17. Temperature swing is the total difference between _____.

    A. cut-in and cut-out temperatures
    B. cut-in and the low temperature in a room
    C. cut-out and the high temperature in a room
    D. high and low temperatures in a room

_____ 18. For a heating thermostat, system lag occurs when temperature _____.

    A. drops below the cut-in value
    B. fails to reach the cut-in value
    C. fails to reach the cut-out value
    D. rises above the cut-out value

_____ 19. A heat anticipator's purpose is to switch _____.

    A. on the heating system just before reaching the cut-in temperature
    B. off the blower before turning off heat production
    C. off the compressor before turning off the condenser fan
    D. off the heating system just before reaching the cut-out temperature

_____ 20. A heat anticipator is always installed in _____.

    A. a low-voltage dc circuit
    B. parallel with the thermostat switch
    C. series with the thermostat switch
    D. a three-phase circuit

_____ 21. Anticipators are basically small _____.

    A. capacitors
    B. diodes
    C. resistors
    D. transistor chips

_____ 22. A cooling anticipator's purpose is to switch _____.

    A. on the cooling system just before reaching the cut-in temperature
    B. off the blower before turning off the compressor
    C. off the compressor before turning off the condenser fan
    D. off the compressor just before reaching the cut-out temperature

10

_____ 23. A cooling anticipator is always installed in ____.

    A. a low-voltage dc circuit

    B. parallel with the thermostat switch

    C. series with the thermostat switch

    D. a three-phase circuit

_____ 24. The number of wires to connect to a combination thermostat depends primarily on ____.

    A. how many wires are in the wire bundle at the base plate

    B. technician preference

    C. thermostat features and stages of operation

    D. None of the above.

_____ 25. The primary purpose of a multistage thermostat is that it can control an HVAC system's ____.

    A. capacity

    B. energy usage

    C. length of operation

    D. wear and tear

_____ 26. If a forced-air electric heating system with a multistage thermostat was turned on and sensed that temperature was significantly below set point (more than 20°F or 11°C), it would ____.

    A. energize all of the heating elements

    B. energize only some of the heating elements

    C. operate the burner at full capacity

    D. run the compressor at full speed

_____ 27. If a multistage thermostat signaled for a cooling system's scroll compressor to run at less than full capacity, the scroll compressor would ____.

    A. bypass a small amount of discharge vapor into the suction line

    B. cycle on and off multiple times a minute

    C. unload one or more cylinders

    D. vent a small amount of high-side refrigerant to maintain lower head pressure

_____ 28. A multistage thermostat for a cooling system will have more than one ____ wiring terminal.

    A. G

    B. O/B

    C. W

    D. Y

_____ 29. A multistage thermostat for a heating system will have more than one ____ wiring terminal.

    A. G

    B. O/B

    C. W

    D. Y

36.3 Line-Voltage Thermostats

_____ 30. In a single-family residence, line voltage is considered ____.

    A. 5 Vdc

    B. 24 Vac

    C. 120 Vac or 240 Vac

    D. 480 Vac

Name _____

_____ 31. In a single-family residence, a line-voltage thermostat is primarily used to control ____.

A. electric baseboard heaters
B. an electric furnace
C. a gas furnace
D. a heat pump

_____ 32. A line-voltage thermostat ____.

A. can conduct higher current than other types of thermostats
B. must be installed in a locked box that is not accessible to the general public
C. must control devices through an isolation relay
D. requires the use of a step-down transformer

36.4 Low-Voltage Thermostats

_____ 33. A low-voltage thermostat is typically supplied with ____.

A. 5 Vdc
B. 24 Vac
C. 120 Vac or 240 Vac
D. 480 Vac

_____ 34. To control higher-current loads, a low-voltage thermostat will usually use ____.

A. an isolation transformer
B. a knife switch
C. relays and contactors
D. a thermistor

_____ 35. The VA rating of step-down transformers used with thermostats is used to determine ____.

A. primary side current rating
B. primary side voltage
C. secondary side current rating
D. secondary side voltage

_____ 36. Low-voltage thermostat wiring for lengths up to 50′ (15 m) should normally be at least ____.

A. 10 AWG
B. 12 AWG
C. 16 AWG
D. 18 AWG

_____ 37. Low-voltage thermostat wiring for lengths over 50′ (15 m) should normally be at least ____.

A. 10 AWG
B. 12 AWG
C. 16 AWG
D. 18 AWG

10

36.5 Millivolt Thermostats

_____ 38. Millivolt thermostats can only be used in certain _____ systems.
   A. air conditioning
   B. commercial refrigeration
   C. heating
   D. heat pump

_____ 39. An advantage that millivolt thermostats have over other thermostats is that they require no _____.
   A. external power source
   B. hardware
   C. wiring
   D. All of the above.

_____ 40. The sensing element used in millivolt thermostats is a(n) _____.
   A. hygroscopic element
   B. sensing bulb and bellows
   C. thermistor
   D. thermocouple

_____ 41. To operate, a millivolt thermostat depends on a(n) _____.
   A. constant, steady draft
   B. external power source
   C. secondary heat exchanger
   D. standing pilot light

_____ 42. The electrical load that a millivolt thermostat is used to actuate is a _____.
   A. condensing unit contactor
   B. reversing valve
   C. sequencer
   D. solenoid valve

36.6 Digital and Programmable Thermostats

_____ 43. *True or False?* The ability of modern thermostats to interface with mobile devices (smartphones, laptops, etc.) allows users to remotely control indoor environments.

_____ 44. A thermostat that draws a small amount of power from the low-voltage side of the HVAC system control to operate its electronics is a _____ thermostat.
   A. multistage
   B. power-stealing
   C. wireless
   D. zoning

_____ 45. Being able to schedule automatic operation using user set points is an advantage of a _____.
   A. multistage
   B. power-stealing
   C. programmable
   D. wireless

Name _____

_____ 46. To prevent relay activation and other circuit board problems, power-stealing thermostats often require the addition of a bypass _____ to the circuit.

    A. capacitor

    B. manual switch

    C. resistor

    D. thermistor

36.7 Thermostat Installation

Match each of the thermostat wire terminal designations with its proper term.

_____ 47. Common connection		A. Aux/E
_____ 48. Compressor contactor (stage 2)		B. C
_____ 49. Condensing unit (compressor) contactor		C. DHD
_____ 50. Cooling power		D. G
_____ 51. Damper		E. HMD
_____ 52. Dehumidifier		F. In or IDT
_____ 53. Fan relay (cooling only)		G. M1 of DMP
_____ 54. Heating power		H. O/B
_____ 55. Heat pump auxiliary heat		I. Rc
_____ 56. Heat relay (stage 1)		J. Rh
_____ 57. Heat relay (stage 2)		K. W
_____ 58. Humidifier		L. W2
_____ 59. Indoor temperature sensor		M. Y
_____ 60. Reversing valve		N. Y2

_____ 61. HVAC system settings are often set and managed using _____ on a circuit board or thermostat subbase.

    A. DIP switches

    B. wiring harnesses

    C. wiring terminals

    D. None of the above.

_____ 62. Free cooling is a method of cooling a building using a(n) _____.

    A. air-side economizer

    B. attic fan

    C. earth-sheltered outer walls

    D. evaporation of potable water

10

_____ 63. Free cooling is best done when outdoor air is _____.
   A. cool and dry
   B. cool and humid
   C. hot and dry
   D. hot and humid

_____ 64. For best results, free cooling should be used on a thermostat that has _____.
   A. enthalpy control
   B. power-stealing capabilities
   C. solid-state switching
   D. wireless connectivity

_____ 65. When free cooling is used on a forced-air system with mechanical air conditioning, a _____ thermostat must be used.
   A. learning
   B. millivolt
   C. multistage
   D. wireless

Match each of the letters in the diagram with the name of its system component.

_____ 66. Air handler

_____ 67. Barometric damper (open)

_____ 68. Inlet damper (open)

_____ 69. Return damper (closed)

_____ 70. Return plenum

_____ 71. Supply plenum

Goodheart-Willcox Publisher

_____ 72. When installing a wireless thermostat, a good location for a temperature sensor would be _____.
   A. anywhere on the ceiling
   B. behind a door or in a corner
   C. directly in front of a supply air register
   D. on an inner wall about 5′ above the floor

Name _____

36.8 Thermostat Diagnostics

_____ 73. When an HVAC system is set to operate in heating mode, a meter reading should indicate a(n) _____ across terminals R and Y.

    A. closed circuit
    B. high capacitance
    C. low current
    D. open circuit

_____ 74. When an HVAC system is operating in cooling mode, a meter reading should indicate a(n) _____ across terminals R and W.

    A. closed circuit
    B. high capacitance
    C. low current
    D. open circuit

_____ 75. When troubleshooting an HVAC system, if no voltage is measured across any of the pairs of thermostat wiring terminals, the problem is most likely with the wire connected to terminal _____.

    A. G
    B. R
    C. W
    D. Y

_____ 76. When troubleshooting a heating system that does not work, if source voltage can be measured in the thermostat, the problem is most likely with the wire connected to terminal _____.

    A. G
    B. R
    C. W
    D. Y

_____ 77. When measuring voltage across an open switch on a thermostat, expect to read _____.

    A. 0 V
    B. source voltage
    C. OL
    D. None of the above.

_____ 78. If a heating system turns on once jumpers are placed across the proper wiring terminals in the thermostat, the problem is most likely the _____.

    A. heating unit
    B. thermostat
    C. wires between thermostat and heating unit
    D. None of the above.

10

_____ 79. When troubleshooting a cooling system that does not work, if source voltage can be measured in the thermostat, the problem is most likely with the wire connected to terminal _____.

A. G
B. R
C. W
D. Y

_____ 80. A multistage thermostat will only close its switch to terminal _____ once temperature drops below set point by more than 3°F (1.5°C).

A. W
B. W2
C. Y
D. Y2

_____ 81. A multistage thermostat will only close its switch to terminal _____ once temperature rises above set point by more than 3°F (1.5°C).

A. W
B. W2
C. Y
D. Y2

_____ 82. A step-down transformer used with a low-voltage thermostat should _not_ have _____.

A. any electrical loads in parallel
B. any electrical loads in series
C. any type of circuit protection devices
D. only one electrical load

_____ 83. To compare a step-down transformer's secondary side VA rating with the VA of all the electrical loads in its circuit, a technician must add together each load's _____.

A. current rating
B. voltage rating
C. resistance
D. All of the above.

_____ 84. In residential applications, live electrical measurements across terminals in a line-voltage thermostat should be _____.

A. 5 Vdc
B. 24 Vac
C. 120 Vac or 240 Vac
D. 480 Vac

_____ 85. If line voltage is _not_ measured across any of the terminals on the subbase of a line-voltage thermostat, then _____.

A. replace only the subbase
B. replace the thermostat with a new line-voltage thermostat
C. replace the thermostat with a new low-voltage thermostat
D. the problem is in the wiring

Name _____

36.9 Zoned Systems

_____ 86. A zoned electric baseboard heating system provides zoning by controlling _____.

    A. dampers
    B. electrical supply
    C. valves and pumps
    D. All of the above.

_____ 87. A zoned forced-air HVAC system provides zoning primarily by controlling _____.

    A. dampers
    B. electrical supply
    C. valves and pumps
    D. All of the above.

_____ 88. A zoned hydronic system provides zoning by controlling _____.

    A. dampers
    B. electrical supply
    C. valves and pumps
    D. All of the above.

_____ 89. If there is increased air pressure on the blower wheel of a forced-air zoned system, then the blower _____.

    A. control board will burn up
    B. fan blades will bend out of shape
    C. motor will draw more current than normal
    D. wheel mounting will loosen

_____ 90. When a forced-air zoned system that controls three zones receives a call to condition only one zone, sufficient airflow through the air handler is most easily, efficiently, and cost-effectively maintained by use of _____.

    A. bypass dampers
    B. multiple air handlers
    C. venting conditioned air to outdoors
    D. All of the above.

_____ 91. Most rectangular dampers used to control airflow to a zone are _____ dampers.

    A. barometric
    B. multiple-blade
    C. single-blade
    D. temperature-operated

_____ 92. In order to close a normally open, two-wire zone damper, _____.

    A. remove its return spring
    B. switch off power to its motor
    C. switch on power to its motor
    D. turn off the blower to reduce static air pressure

10

_____ 93. An easy way to identify a power-open/power-closed damper is that it will have _____ connected to its motor.

    A. one wire
    B. two wires
    C. three wires
    D. two springs

_____ 94. A power-open/power-closed damper uses a(n) _____ to determine when the damper is in the closed position.

    A. end switch
    B. proximity switch
    C. push button
    D. set of normally open contacts along its sealing edge

_____ 95. A barometric damper's response to static air pressure can be accurately changed by adjusting _____.

    A. the dial on its pressure controller
    B. its settings on the zone control panel
    C. its spring
    D. its weights

_____ 96. In a zoned forced-air system with mechanical cooling, a freezestat opens with the purpose of turning off the _____.

    A. air-side economizer
    B. blower
    C. compressor
    D. zone control panel

_____ 97. If a small amount of frost begins to build up on an evaporator due to lower than normal airflow rates in a forced-air zoned system that is feeding air to only two of its three zones, the best solution from a zone control panel would be to _____.

    A. automatically cycle off the compressor
    B. open a motorized bypass damper and continue cooling operation
    C. reverse compressor motor rotation to reverse refrigerant flow
    D. turn on electric defrost elements and continue operation

_____ 98. If a solid ice buildup blocks an entire evaporator causing lower than normal airflow rates in a forced-air zoned system that is trying to feed air to only two of its three zones, the best solution from a zone control panel would be to _____.

    A. automatically cycle off the compressor
    B. open a motorized bypass damper and continue cooling operation
    C. reverse compressor motor rotation to reverse refrigerant flow
    D. turn on electric defrost elements and continue operation

_____ 99. Once all thermostats are satisfied in a forced-air zoned system, operation usually ends with all dampers _____.

    A. fully open
    B. at the halfway position
    C. 3/4 closed
    D. fully closed

Name _____

_____ 100. When measuring the electrical resistance of a damper motor in a forced-air zoned system, a good motor should have _____.

    A. 0 Ω

    B. a small resistance value

    C. several megohms of resistance

    D. infinite resistance

Critical Thinking

101. Explain why electrical contact point bounce is undesirable. Explain how it can be reduced or prevented.

_____

_____

_____

_____

_____

102. Explain how the temperature differential setting is adjusted on older mechanical heating thermostats.

_____

_____

_____

_____

103. Explain why millivolt thermostats have been phased out.

_____

_____

_____

_____

_____

104. List several factors that may affect a thermostat and require that its location be changed.

_____

_____

_____

105. Explain why "free cooling" operation is considered free.

_____

_____

_____

_____

106. A customer with a forced-air heating and cooling system asks you to equip the system to operate in free cooling mode. It already has a thermostat that can be wired for free cooling operation. What other parts and components will you need to purchase and install? Explain how you will modify the existing wiring and ductwork. Explain how and where you will install each of the new parts.

_____

_____

_____

_____

_____

_____

_____

_____

_____

107. Each step-down transformer used with a low-voltage thermostat has a VA rating on its secondary side. The electrical loads connected to a step-down transformer's secondary side have a combined VA value that can be calculated. Should the total VA of the electrical loads be less than, equal to, or above the transformer's secondary side VA rating? Explain your answer.

_____

_____

_____

_____

_____

_____

Heating and Cooling Loads

Carefully study the chapter and then answer the following questions.

37.1 Heat Transfer

_____ 1. The practice of using materials and processes that lower energy costs, reducing operating and maintenance costs, increasing productivity, and decreasing pollution is known as ____ design.
   A. economizer
   B. emissivity
   C. heat lag
   D. sustainable

_____ 2. The increase of heat within a space that results from solar radiation or heat radiated by other sources is referred to as ____.
   A. heat gain
   B. heat loading
   C. heat loss
   D. None of the above.

_____ 3. The transfer of heat from inside to outside by means of conduction, convection, and radiation through all surfaces is known as ____.
   A. heat gain
   B. heat loading
   C. heat loss
   D. None of the above.

_____ 4. *True or False?* During the summer, enough heat must be removed from a conditioned space to cool it as heat enters from the warm outside air.

_____ 5. Designing a building to reduce heat transfer can be described as making a building envelope ____.
   A. loose
   B. open
   C. stamped
   D. tight

10

37.2 Heat Loads

_____ 6. The amount of heat that must be added or removed from a space in order to maintain the desired temperature in that space is referred to as the _____.
A. heat load
B. heat loss
C. system capacity
D. transfer rate

_____ 7. For the purpose of sizing a system, the maximum heat load is determined for a period of _____.
A. 1 minute
B. 1 hour
C. 1/2 day
D. 1 week

_____ 8. An example of a building's latent heat load would be _____.
A. equipment heat from motors and other devices
B. radiant heat from the sun
C. sweating humans and animals
D. None of the above.

37.3 Calculating Heat Leakage

_____ 9. Heat leakage is the heat that is conducted through the walls, ceilings, and floors when there is a(n) _____ between spaces.
A. insulation gap
B. opening
C. temperature difference
D. window

_____ 10. *True or False?* When the temperature outside a conditioned space is lower than the temperature inside the space, heat leakage results in heat gain.

_____ 11. Calculating the total heat leakage for a space requires _____.
A. determining the types of materials through which heat is leaking
B. calculating how much heat is leaking
C. determining the surface area of each type of material
D. All of the above.

_____ 12. A measure of the amount of heat that will pass through one square foot of material one inch thick, in one hour, when there is a 1°F temperature difference between the two sides of a material is called _____.
A. heat capacitance
B. heat transference
C. thermal conductance
D. thermal conductivity

Name _____

_____ 13. A measure of the amount of heat that will pass through one square foot of a component (any thickness) in one hour when there is a temperature difference of 1°F between one side of the component and the other is called _____.

    A. heat capacitance
    B. heat transference
    C. thermal conductance
    D. thermal conductivity

_____ 14. A measure of a material's resistance to heat transfer is known as its _____.

    A. leak resistance
    B. heat capacitance
    C. thermal conductance
    D. thermal resistance

_____ 15. The U-value of a material is its thermal _____.

    A. conductance
    B. conductivity
    C. resistance
    D. transmittance

_____ 16. The K-value of a material is its thermal _____.

    A. conductance
    B. conductivity
    C. resistance
    D. transmittance

_____ 17. *True or False?* The higher a material's R-value, the slower the rate at which heat will transfer through that material or component.

_____ 18. *True or False?* Building boundary air films add extra insulation between the building component and the surrounding environment because they have a slightly different temperature than the surrounding air.

_____ 19. C-values, K-values, and U-values as recorded are only accurate when the outside air is exactly 1°F warmer or cooler than the inside air. To account for a greater temperature difference, multiply these values by the _____.

    A. actual temperature difference
    B. surface area
    C. thickness of the material
    D. None of the above.

_____ 20. The ideal indoor temperature which the heating or cooling system works to maintain is the indoor _____ temperature.

    A. comfort
    B. design
    C. ideal
    D. standard

10

_____ 21. The extremely cold conditions for a geographical region where a heating system is being installed is known as the _____ temperature.

    A. climate design
    B. outdoor design
    C. outdoor ideal
    D. standard climate

_____ 22. Regarding the three values supplied on ASHRAE charts for heating system sizing, the higher the percentage number, _____.

    A. the greater the need for a higher capacity heating system
    B. the less time the area will spend above the listed temperature
    C. the more time the area will spend above the listed temperature
    D. the more time the area will spend below the listed temperature

_____ 23. When calculating heat gain for the purpose of sizing a cooling system, technicians typically use _____ for the summertime design temperature.

    A. 50°F (13°C) at 85% rh
    B. 65°F (18°C) at 75% rh
    C. 75°F (24°C) at 50% rh
    D. 85°F (30°C) at 5% rh

_____ 24. The term *heat transfer multiplier (HTM)* refers to a building component's _____ multiplied by a temperature difference.

    A. C-value
    B. K-value
    C. R-value
    D. U-value

_____ 25. If the exterior dimensions of a building's walls are used to calculate the surface area of the walls, the dimensions will result in _____ heat leakage loads than if the inside dimensions are used.

    A. much higher
    B. slightly higher
    C. slightly lower
    D. the same

_____ 26. An additional framed pane of glass that can be attached to the outside of the window to provide additional insulation is a(n) _____.

    A. anti–heat leak pane
    B. extra glazing pane
    C. sandwich frame
    D. storm sash

_____ 27. *True or False?* Heat transfer rate is the amount of heat conducted through a structure for a given amount of time.

_____ 28. Heat transfer rate can be calculated using a building's _____.

    A. C-values
    B. K-values
    C. U-values or R-values
    D. All of the above.

Name _____

37.4 Other Factors Affecting Heat Loads

_____ 29. Basement heat loss is generally calculated using _____.
  A. area of the basement floor and walls
  B. basement building material and insulation U-values
  C. the temperature difference between the basement and the ground
  D. All of the above.

_____ 30. Plastic sheeting between soil and basement walls primarily functions _____.
  A. to increase the insulation value of the wall
  B. to keep the construction site clean
  C. as a vapor barrier
  D. None of the above.

_____ 31. If the air pressure inside a building is lower than the air pressure outside the building, air will _____.
  A. leak into the building
  B. leak out of the building
  C. remain stationary indoors and outdoors
  D. None of the above.

_____ 32. If the air pressure inside the building is greater than the air pressure outside the building, air will _____.
  A. leak into the building
  B. leak out of the building
  C. remain stationary indoors and outdoors
  D. None of the above.

_____ 33. *True or False?* Infiltration calculations are primarily based on the total volume of the building and the length and size of all the cracks in the building.

_____ 34. The ability of light rays to pass through a material is known as _____.
  A. emissivity
  B. luminescence
  C. reflectivity
  D. transference

_____ 35. The amount of time it takes for heat to travel through a substance that is heated on one side is known as _____.
  A. heat lag
  B. system lag
  C. system overshoot
  D. thermal transit time

_____ 36. *True or False?* When the sun heats an outside wall, several hours may pass before the heat reaches the inner surfaces of the wall.

_____ 37. Latent heat in a building increases as air _____.
  A. moisture content decreases
  B. moisture content increases
  C. temperature decreases
  D. temperature increases

10

37.5 Heating and Cooling Load—Manual J Method

_____ 38. The organization that publishes standardized forms and software for calculating heating and cooling loads of buildings is ____.
   A. ACCA
   B. ASHRAE
   C. NATE
   D. RSES

_____ 39. A technician that is calculating heat loads for small commercial buildings should reference Manual ____.
   A. D
   B. J
   C. N
   D. S

_____ 40. A technician that is calculating heat loads for residential buildings should reference Manual ____.
   A. D
   B. J
   C. N
   D. S

_____ 41. Air leakage is quantified in ____.
   A. ACH
   B. Btu/hr
   C. cfm
   D. tons of refrigeration

_____ 42. _True or False?_ Manual J views any type of window without distinction and does not take into account whether the window is in a shaded area.

37.6 Software and Apps for Load Calculations

_____ 43. Full, thorough, and in-depth Manual J calculations are usually performed by ____.
   A. an average HVAC technician
   B. a first-year apprentice
   C. the homeowner
   D. technicians who do load calculations full time

_____ 44. _True or False?_ The only Manual J software that can be legally used for contracted work is produced and released by ACCA.

Name _____

Critical Thinking

45. What can happen if a building envelope is made too tight? What can be done to fix this?

_____

_____

_____

_____

46. Explain how the net wall area is computed for a building.

_____

_____

_____

47. List three or more methods that can be used to reduce the cooling loads in a building.

_____

_____

_____

10

Notes

CHAPTER 38

Forced-Air Heating Fundamentals

Name _____

Date _____ Class _____

> *Carefully study the chapter and then answer the following questions.*

38.1 Basic Components

_____ 1. Heat distribution for conditioned spaces in HVAC systems is primarily done through the following methods, *except* _____.

    A. conduction
    B. forced-air
    C. hydronic
    D. radiant

_____ 2. In a forced-air system, the component responsible for delivering heated air to the conditioned space is the _____.

    A. blower
    B. combustion blower
    C. flue
    D. heat exchanger

_____ 3. Before forced-air systems were developed, gravity heating systems relied on _____ to move air to and from conditioned spaces.

    A. conduction
    B. gravity
    C. natural frequencies
    D. radiation

_____ 4. A gravity heating system had to be installed _____ for the system to be effective.

    A. in a lower part of a building
    B. outside
    C. on the top floor of a building
    D. in the very center of a building

_____ 5. An electric heating system _____.

    A. uses one or more heat exchangers
    B. uses only one combustion blower
    C. has one or more resistance heating elements
    D. has a flue or vented outlet

11

_____   6. A furnace's heat exchanger is where heat is transferred from ____.

    A. circulating air to the evaporator

    B. combustion gases or electric elements to circulating indoor air

    C. the condenser to outdoor air

    D. the liquid line to the suction line

_____   7. A secondary heat exchanger differs from a primary heat exchanger in that it ____.

    A. connects in parallel with the primary heat exchanger

    B. separates the primary heat exchanger from the burners

    C. transfers both sensible and latent heat

    D. None of the above.

Identify the components of the high-efficiency furnace shown in the following image.

_____   8. Burners

_____   9. Combustion blower

_____   10. Indoor blower

_____   11. Primary heat exchanger

_____   12. Secondary heat exchanger

Rheem Manufacturing Company

Name _____

_____ 13. Due to system operation, a furnace's secondary heat exchanger may be made of _____.

    A. cadmium

    B. masonry

    C. PVC

    D. stainless steel

_____ 14. Due to system operation, a condensing furnace's exhaust piping is often made of _____.

    A. cadmium

    B. masonry

    C. PVC

    D. stainless steel

_____ 15. A condensing furnace has an annual fuel utilization efficiency (AFUE) rating that is _____.

    A. 50% to 65%

    B. 70% to 75%

    C. 80% to 90%

    D. above 90%

_____ 16. *True or False?* Condensing furnace combustion gases must be vented through a chimney.

_____ 17. An indoor blower with a direct-drive motor will run at _____.

    A. double the motor speed

    B. half the motor speed

    C. the same speed as the motor

    D. None of the above.

_____ 18. For belt-driven indoor blowers, as the size of the pulley on the motor increases, the blower speed _____.

    A. decreases

    B. increases

    C. remains the same

    D. None of the above.

_____ 19. For belt-driven indoor blowers, as the size of the pulley on the blower increases, the blower speed _____.

    A. decreases

    B. increases

    C. remains the same

    D. None of the above.

11

Identify the components of the belt-driven blower shown in the following image.

_____ 20. Blower fan wheel

_____ 21. Blower housing

_____ 22. Blower pulley

_____ 23. Motor

_____ 24. Motor pulley

_____ 25. V-belt

Goodheart-Willcox Publisher

_____ 26. *True or False?* Belt-driven blowers are inherently more efficient than direct-drive blowers.

_____ 27. A combustion blower moves fresh air and combustion air through the _____.

    A. conditioned space

    B. heat exchanger

    C. return ducts

    D. supply ducts

38.2 Furnace Types and Construction

_____ 28. A furnace that is adjustable so that a technician can configure it as upflow, downflow, or horizontal is a _____ furnace.

    A. makeup air

    B. modulating

    C. multipoise

    D. two-stage

_____ 29. A highboy furnace is _____.

    A. arranged inside to minimize the height of its cabinet

    B. installed in areas where vertical space is not an issue

    C. longer or wider than most furnaces

    D. All of the above.

_____ 30. A lowboy furnace is _____.

    A. arranged inside to minimize the height of its cabinet

    B. installed in areas where vertical space is not an issue

    C. taller than most furnaces

    D. All of the above.

Name _____

_____ 31. An upflow furnace draws in air from its _____.

    A. lower portion

    B. middle portion

    C. upper portion

    D. None of the above.

_____ 32. A downflow furnace draws in air from its _____.

    A. lower portion

    B. middle portion

    C. upper portion

    D. None of the above.

_____ 33. A furnace is considered two-stage because it can _____.

    A. be arranged taller or shorter to fit different spaces

    B. be installed in any position

    C. be wired for 120 Vac or 240 Vac

    D. produce two different levels of heat

_____ 34. A furnace is considered modulating because it can _____.

    A. be arranged taller or shorter to fit different spaces

    B. be installed in any position

    C. heat and cool at the same time

    D. vary its heat output from 40% to 100% of its total capacity

38.3 Forced-Air Duct Arrangements

_____ 35. A forced-air system usually contains an air filter _____.

    A. between the blower and the evaporator

    B. between the heat exchanger and plenum

    C. on the return air side of the system

    D. on the supply air side of the system

_____ 36. Air that has just been heated flows through a forced-air system's _____ ducts.

    A. exhaust air

    B. return

    C. supply

    D. None of the above.

_____ 37. Air that is about to be heated flows through a forced-air system's _____ ducts.

    A. fresh air inlet

    B. return

    C. supply

    D. None of the above.

11

38.4 Makeup Air Units

_____ 38. *True or False?* Negative pressure in a building can cause exfiltration around doors and windows.

_____ 39. Building codes may require the use of makeup air units for the purpose of preventing _____ from developing in a house.
   A. CO concentrations
   B. CO_2 concentrations
   C. negative pressure
   D. positive pressure

_____ 40. Makeup air is commonly deposited directly into the _____.
   A. garage
   B. kitchen
   C. return air duct
   D. supply air duct

38.5 Blower Controls

_____ 41. *True or False?* A combustion-type furnace's burners and blower both turn off and on at the same time.

_____ 42. A time-delay control on a circuit board is often adjusted using a set of _____.
   A. DIP switches
   B. knife switches
   C. pressure switches
   D. relays

_____ 43. A thermostatic control may delay turning on a blower until the plenum has reached _____.
   A. 60°F (15.5°C)
   B. 90°F (32°C)
   C. 140°F (60°C)
   D. 300°F (149°C)

_____ 44. A thermostatic control may delay turning off a blower until the plenum has reached _____.
   A. 60°F (15.5°C)
   B. 90°F (32°C)
   C. 140°F (60°C)
   D. 300°F (149°C)

_____ 45. In a heating system, a limit control is a temperature-sensing switch that _____ as long as the plenum temperature remains below its set point.
   A. oscillates at a regular frequency
   B. remains closed
   C. remains open
   D. None of the above.

Name _____

_____ 46. When a plenum limit control actuates, the most likely and direct cause would be a _____.

  A. combustion blower running in reverse
  B. blower not blowing
  C. burner not burning
  D. massive heat exchanger leak

38.6 Unit Heaters

_____ 47. Unit heaters commonly produce heat from the following, *except* _____.

  A. coal
  B. electric
  C. gas
  D. oil

_____ 48. Unit heaters are most often used to warm _____.

  A. each of the apartments in a large complex
  B. an entire home with zoned control
  C. large, open areas
  D. small additions to houses

_____ 49. In summer, the heating section of a unit heater can be turned off and high flow air can be directed across doorways and building openings to function as a(n) _____.

  A. air curtain
  B. air-side economizer
  C. indoor blower
  D. makeup air unit

Critical Thinking

50. Explain why there is a difference in efficiency between a noncondensing furnace and a condensing furnace. How is this difference achieved?

_____

_____

_____

_____

_____

51. Explain what will happen (in short-term and in long-term) if a blower's fan blades become bent or broken.

_____

_____

_____

11

52. A two-stage furnace has two stages of operation. Describe when the furnace would operate in each of these stages. Explain the advantages of each stage of operation.

_____

_____

_____

_____

53. Explain how a makeup air unit affects the operation of a combustion heating system. What would happen if the makeup air unit did not perform its function?

_____

_____

_____

_____

_____

54. Explain why a time-delay control is used to turn on a furnace's indoor blower.

_____

_____

39

Hydronic Heating Fundamentals

Name _____

Date _____ Class _____

Carefully study the chapter and then answer the following questions.

39.1 Hydronic System Components

_____ 1. A conventional boiler is designed and should maintain an operational water temperature of _____.
   A. 90°F (32°C)
   B. 140°F (60°C)
   C. 200°F (93°C)
   D. 350°F (177°C)

_____ 2. To avoid damage, condensing boilers are generally equipped with a secondary heat exchanger that is made of _____.
   A. aluminum
   B. cast iron
   C. copper
   D. stainless steel

_____ 3. A dry-base boiler is dry in the _____.
   A. area under the combustion chamber
   B. flue
   C. radiators
   D. All of the above.

_____ 4. Boilers that are lighter and classified as low-mass, low-volume are generally made of _____.
   A. cast iron
   B. cement
   C. copper
   D. steel

_____ 5. *True or False?* Propylene glycol is less environmentally friendly than ethylene glycol.

_____ 6. One way to prevent corrosion is to add _____ to the water, which interact chemically with contaminates, rendering them inert.
   A. chemical scavengers
   B. corrosion inhibitors
   C. deaerators
   D. embrittlers

11

_____ 7. Another way to prevent corrosion is to add _____ to the water, which adds a protective coating to the surface of the metal.

    A. chemical scavengers
    B. corrosion inhibitors
    C. deaerators
    D. embrittlers

_____ 8. A weakening of metal caused by long-term corrosion is known as _____.

    A. cavitation
    B. deaeration
    C. embrittlement
    D. oxidation

_____ 9. The proper pH range for water in a hydronic heating system is _____.

    A. between 1 and 5
    B. 6
    C. between 7 and 10
    D. over 10

_____ 10. The part of a circulating pump that moves water through a hydronic system is the _____.

    A. impeller
    B. terminal box
    C. windings
    D. None of the above.

Match the components of the older-style boiler system shown with the correct term.

_____ 11. Impeller

_____ 12. Motor

_____ 13. Motor shaft

_____ 14. Terminal box

_____ 15. Water return (inlet)

_____ 16. Water supply (outlet)

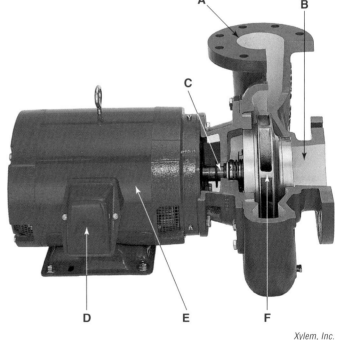

Xylem, Inc.

Name _____

_____ 17. One side of an expansion tank is filled with water and connected to the hydronic system piping and the other side of the tank is filled with _____.
  A. air
  B. oil
  C. refrigerant
  D. sand

_____ 18. When charging an expansion tank, a technician should use a _____.
  A. bicycle pump
  B. circulating pump
  C. pressurized cylinder
  D. vacuum pump

_____ 19. The purpose of an expansion tank is to maintain a stable _____ under varying operating conditions.
  A. heat transfer
  B. pressure
  C. temperature
  D. water pH

_____ 20. *True or False?* An expansion tank must be installed on the supply side of system piping and never on the return side.

11

> *Match the components of the older-style boiler system shown with the correct term.*

_____ 21. Backflow preventer

_____ 22. Boiler fitting

_____ 23. Circulating pump

_____ 24. Combination valve

_____ 25. Expansion tank

_____ 26. Flow-control valve

_____ 27. Pressure-reducing valve

_____ 28. Pressure-relief valve

_____ 29. Water return

_____ 30. Water supply

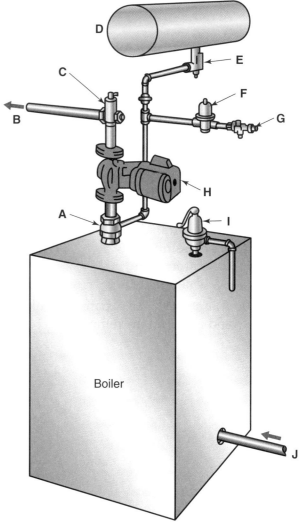

Goodheart-Willcox Publisher

_____ 31. A _____ valve blends hot water from one inlet with cooler water from another inlet.
- A. backflow
- B. balancing
- C. check
- D. mixing

_____ 32. Balancing valves are most often _____ valves.
- A. butterfly
- B. diaphragm
- C. globe or ball
- D. needle

Name _____

_____ 33. If the supply water leaving a boiler is *not* hot enough, the water will be _____.

A. directed into an aftermarket flue gas reclamation heat exchanger
B. drained away
C. pumped into the expansion tank
D. recirculated through the boiler

_____ 34. A mixing valve may be installed _____.

A. at the overflow valve outlet
B. in the return or supply water line
C. only in the return water line
D. only in the supply water line

_____ 35. A valve that adjusts the water flow to each terminal unit or zone so that heat is evenly distributed is called a _____.

A. balancing valve
B. thermostatic mixing valve
C. mixing valve
D. pressure-reducing valve

_____ 36. A valve installed in the makeup water line that adjusts the pressure of the water supply from the main to the operating pressure of the system is a _____ valve.

A. balancing
B. mixing
C. pressure-reducing
D. pressure relief

_____ 37. To prevent water migration during the Off cycle, passageways are closed using _____ valves.

A. manual needle
B. pressure-actuated butterfly
C. thermostatic solenoid
D. weighted check

_____ 38. A specialized check valve installed in the makeup water line that prevents water in a hydronic system from flowing back into the water main is a _____.

A. backflow preventer
B. balancing valve
C. pressure-reducing valve
D. pressure-relief valve

_____ 39. Heat exchangers that transfer heat from the circulating water in a hydronic system to the air in a conditioned space are called _____.

A. boilers
B. heat impellers
C. terminal units
D. None of the above.

_____ 40. *True or False?* Radiators provide heat to a conditioned space only through radiation.

_____ 41. A terminal unit equipped with a fan that blows air across the surface of the heating element is a _____.

    A. fan convector
    B. heat impeller
    C. radiator
    D. All of the above.

_____ 42. The type of air that consists of small air bubbles that travel along with the circulating water is called _____.

    A. dissolved air
    B. entrained air
    C. free air
    D. vent air

_____ 43. The type of air that consists of bubbles at the high points of the hydronic system is called _____.

    A. dissolved air
    B. entrained air
    C. free air
    D. vent air

_____ 44. Water's ability to hold dissolved oxygen increases as _____.

    A. its pressure drops
    B. its temperature drops
    C. its temperature rises
    D. None of the above.

_____ 45. An air vent releases air to atmosphere by means of a _____.

    A. float valve
    B. pressure valve
    C. solenoid valve
    D. temperature valve

_____ 46. Devices that use a series of deflectors to collect and remove air from water in hydronic systems are _____.

    A. air separators
    B. air scoops
    C. air valves
    D. None of the above.

_____ 47. Devices that use a wire mesh element to create a swirling motion to remove air from water in hydronic systems are _____.

    A. air separators
    B. air scoops
    C. air valves
    D. None of the above.

Name _____

> *Match the boiler components shown with the correct term.*

_____ 48. Air separator

_____ 49. Backflow preventer

_____ 50. Connected to makeup water line

_____ 51. Expansion tank

_____ 52. Pressure-reducing valve

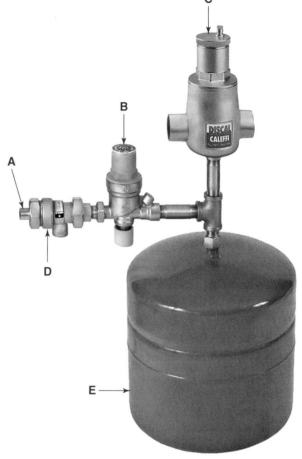

© 2012 Caleffi North America, Inc.

39.2 Hydronic System Designs

_____ 53. Splitting a series loop hydronic system into two smaller loops _____.
A. increases the number of components that the circulating water must pass through
B. increases the overall distance that the circulating water must travel
C. reduces the overall distance that the circulating water must travel
D. None of the above.

_____ 54. *True or False?* A major difference between a series loop and a one-pipe system is that a one-pipe system has branch circuits.

11

Match the letter of each hydronic piping arrangement shown with its proper name.

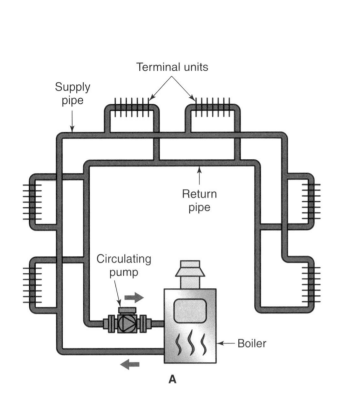

A

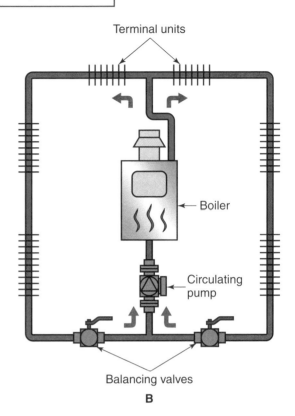

B

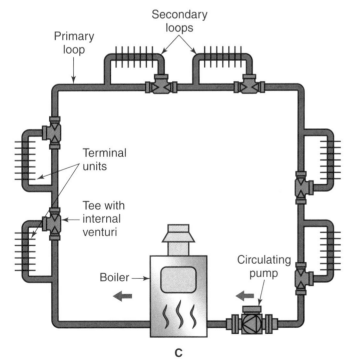

C

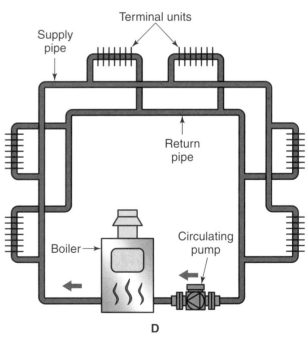

D

_____ 55. Direct return two-pipe system

_____ 56. One-pipe system

_____ 57. Reverse return two-pipe system

_____ 58. Split series loop system

Name _____

_____ 59. In a zoned hydronic system, a zone valve is most often controlled by _____.

    A. the boiler
    B. a circulating pump
    C. a purge valve
    D. a room thermostat

_____ 60. Radiant hydronic systems often use _____ tubing.

    A. bronze
    B. cast iron
    C. galvanized
    D. PEX

_____ 61. *True or False?* In a radiant heating system, the loops of tubing warm the surfaces adjacent to them by conduction.

_____ 62. *True or False?* Dry underfloor radiant heating systems consist of tubing embedded in a concrete slab.

_____ 63. A radiant system traditionally distributes heat to conditioned spaces by _____.

    A. fan convectors
    B. forced air through a centralized ductwork system
    C. piping enclosed in walls or floors
    D. radiators

_____ 64. In an oil-fired boiler, the atomized fuel spray is ignited by _____.

    A. electrical sparks
    B. flint striker
    C. a heated element
    D. pilot light

_____ 65. A constant pressure is maintained in the combustion chamber of an oil-fired boiler by a(n) _____.

    A. automatic draft regulator
    B. pressure-relief valve
    C. stack thermostat
    D. tridicator

_____ 66. Water pressure and temperature in a boiler are measured and displayed on a _____.

    A. high-limit control
    B. low-water cutoff
    C. room thermostat
    D. tridicator

_____ 67. A gas-fired boiler uses a(n) _____ to automatically shut off the gas if the water temperature or pressure gets too high.

    A. automatic draft regulator
    B. high-limit control
    C. stack thermostat
    D. tridicator

11

39.3 Hydronic System Controls

_____ 68. A low-water cutoff is a control that is intended to allow system operation only _____.
   A. after an expansion tank is drained of excess water
   B. when a boiler contains enough water
   C. when boiler water drops below overfill level
   D. None of the above.

_____ 69. Low-water cutoff controls generally actuate based on response to a _____.
   A. crystal or thermocouple
   B. float or sensor probe
   C. flow switch or pressure switch
   D. pressure switch or temperature switch

_____ 70. *True or False?* Excessive air or contaminants in boiler water can cause a low-water cutoff to malfunction.

_____ 71. A boiler's flow switch actuates in reaction to _____.
   A. high flow
   B. inadequate flow
   C. stable, sufficient flow
   D. None of the above.

_____ 72. An aquastat measures the temperature of boiler water and uses this information to _____.
   A. control the zone valves
   B. modulate the mixing valve
   C. turn on and off the circulator pump
   D. turn on and off heating of the water

_____ 73. *True or False?* The maximum temperature setting on an outdoor reset control should be set higher than the temperature setting on the boiler's high-limit control.

_____ 74. Outdoor reset control monitors the difference between indoor and outdoor temperatures in order to modify the _____.
   A. high-limit control value
   B. low-water cutoff setting
   C. operating temperature range of a hydronic system
   D. settings of each room thermostat

_____ 75. An indoor reset control is used to adjust _____.
   A. boiler temperature to match the heating load
   B. circulator pump speed based on room thermostat settings
   C. high-limit control value based on outdoor temperature
   D. room thermostat setting to match outdoor temperature

_____ 76. An indoor or outdoor reset control should be wired to bypass a boiler's high-limit control _____.
   A. in areas with very cold winters
   B. in areas with a very mild climate
   C. in areas with a warm climate
   D. never

Name _____

39.4 Hydronic System Installation

_____ 77. Eccentric reducer fittings are used to join tubing of different sizes in order to _____.

A. increase water pressure on branch circuits

B. introduce a slight turbulence to the flow

C. reduce the danger of air pockets forming

D. save the time it would take to fabricate a solution

_____ 78. When installing metal pipes, be certain to make allowance for _____.

A. damage from heat transferring from hangers to the structure

B. drip leaks by sloping each horizontal run of pipe

C. piping expansion

D. None of the above.

_____ 79. The heat difference between return water and supply water should be approximately _____.

A. 10°F (5°C)

B. 20°F (11°C)

C. 30°F (17°C)

D. 35°F (25°C)

_____ 80. To protect valves and pump seals from dirt, sand, and other particulates when preparing a hydronic system for initial startup, a technician should _____.

A. add two pounds of trisodium phosphate for every 20 gallons of water

B. analyze the water for trisodium phosphate

C. circulate the prepared water solution for about one hour and end installation by beginning system operation

D. clean the screens thoroughly before refilling the system

_____ 81. When balancing a system during initial start-up, pressure gauges or flow meters are used to determine water flow rate in _____.

A. cfm

B. gpm

C. mph

D. None of the above.

_____ 82. For steam heating systems, the _____ must slope down to the boiler.

A. radiator valves

B. return piping

C. supply piping

D. venting piping

_____ 83. A steam trap that uses a mixture of water and mineral spirits to trigger its valve is a(n) _____.

A. balanced pressure steam trap

B. bimetal steam trap

C. expansion steam trap

D. None of the above.

_____ 84. _True or False?_ In a two-pipe system, the thermostatic radiator valve should be installed on the supply side of the radiator.

11

_____ 85. In a one-pipe system, the air vent should be installed on the ____.

    A. boiler

    B. expansion tank

    C. fill valve

    D. thermostatic radiator valve

39.5 Troubleshooting and Servicing Hydronic Systems

_____ 86. Blockage in the intake and exhaust passages can cause ____.

    A. the high-limit control to trip

    B. incomplete combustion

    C. indoor reset control to malfunction

    D. a terminal unit to become air-bound

_____ 87. A safety control that shuts down the boiler if the burner flames travel beyond the combustion chamber is a ____.

    A. blocked-vent switch

    B. combustion switch

    C. high-limit control

    D. rollout switch

_____ 88. A safety control that shuts down the boiler if a backdraft condition is detected is a ____.

    A. blocked-vent switch

    B. combustion switch

    C. high-limit control

    D. rollout switch

_____ 89. The cause of most water circulation problems in a hydronic heating system is ____.

    A. air trapped in the system

    B. circulator pump motor burnout

    C. control valves stuck closed

    D. incorrect sizing of tubes or pipes

_____ 90. The damage caused to circulator pumps through the implosion of bubbles in hydronic system water is called ____.

    A. cavitation

    B. free air corrosion

    C. impeller bubblation

    D. water hammer

_____ 91. A component in a hydronic heating system is ____ if enough air gathers at that location to completely block the flow of water.

    A. air-bound

    B. air-entrained

    C. cavitated

    D. deaerated

Name _____

_____ 92. The most likely cause of a buildup of air in a system equipped with air vents is _____.

A. malfunctioning air vents
B. a puncture in the expansion tank element
C. slow running of the circulator pump
D. None of the above.

_____ 93. Purge valves combine the function of a(n) _____ in a single valve body.

A. air vent and a drain valve
B. backflow preventer and a drain valve
C. drain valve and a mixing valve
D. drain valve and a shutoff valve

_____ 94. The most likely consequence of a system with an expansion tank that is waterlogged is the _____.

A. backflow preventer allowing system water to flow back into the mains
B. boiler exploding
C. pressure-reducing valve allowing more water into the system
D. pressure-relief valve opening to lower system pressure

_____ 95. Which of the following should *not* be done when recharging a diaphragm expansion tank?

A. Check the pressure every two or three strokes of the pump.
B. Leave the makeup water line open.
C. Use a bicycle pump to increase air pressure if necessary.
D. Use a pressure gauge to measure the air pressure in the tank.

_____ 96. If a hydronic system is providing uneven heating, a technician should check _____.

A. for air blockages
B. the operation of circulating pumps and zone valves
C. the operation of thermostatic controls
D. All of the above.

_____ 97. Which of the following is true when checking a steam heating system?

A. Check the pressure gauge after draining the boiler water.
B. If no water shows in the water level sight glass, shut the system off at once.
C. If the water level is one-half full, fill it up to 100%.
D. The water level should show the boiler water level at 85% to 100% full.

_____ 98. When servicing a steam heating system, a technician should _____.

A. briefly pull the lever on the pressure-relief valve to make sure that it is not stuck
B. ensure the pressure-relief valve is equipped with a discharge tube
C. immediately shut the system off if no water or steam comes out when the pressure-relief valve is opened
D. All of the above.

_____ 99. If a radiator in a steam heating system is air-bound, the _____.

A. air vent is not working
B. steam trap is not working
C. thermostat for the radiator valve is not working
D. water level is too low

11

39.6 Preparing a Boiler for the Heating Season

_____ 100. Which of the following is *not* a step in inspection and maintenance procedures for an oil-fired boiler?

    A. Check and tighten all electrical connections.
    B. Clean and adjust the ignition electrodes.
    C. File out the orifice of the oil burner nozzle.
    D. Replace cracked or damaged insulation.

_____ 101. If a blower motor is found to be warmer than normal when inspecting a system, new _____ may be necessary.

    A. bearings
    B. cooling fins
    C. mounting hardware
    D. None of the above.

_____ 102. Local building codes typically require that a boiler system pass a post-installation _____ test.

    A. customer satisfaction
    B. pressure
    C. temperature
    D. time

Critical Thinking

103. Explain why temperatures remain steady longer in hydronic systems than in forced-air systems.

_____

_____

_____

104. When the water in a conventional boiler drops below 140°F (60°C) and the system continues operating for some time below that temperature, explain what happens and also what could happen later.

_____

_____

_____

105. Explain why corrosion occurs inside hydronic piping.

_____

_____

106. Explain what a wet-rotor centrifugal pump is. How does it differ from other types of circulating pumps?

_____

_____

_____

Name _____

107. Explain what happens if the two sets of terminal units in a hydronic system are not balanced.

_____

_____

_____

108. Explain why plastic or composite tubing used in a radiant heating system must have an oxygen barrier.

_____

_____

109. Briefly explain how a one-pipe steam system works. How can only one pipe be used?

_____

_____

_____

_____

_____

_____

110. Explain some of the system responses when a boiler's flow switch actuates.

_____

_____

_____

111. Explain which system value changes and how it changes when an outdoor reset control determines that the difference between indoor and outdoor temperatures is decreasing. Explain the purpose for this change.

_____

_____

_____

_____

112. Explain four or more common symptoms of excessive air in a hydronic system.

_____

_____

_____

_____

_____

11

113. When bleeding a system with manual bleed valves, it is a good idea to connect a hose to the valve outlet and place the other end in a bucket filled with several inches of water. Explain the reasons why these steps are done as described.

_____

_____

_____

_____

114. List the three main times when a hydronic system should be purged.

_____

_____

_____

115. Briefly explain how purging differs from bleeding.

_____

_____

116. Explain why a technician should turn off the boiler and allow it to cool down before purging the system.

_____

_____

117. The purging of hydronic systems may differ based on several factors. While many hydronic systems are filled with only water, others are filled with a glycol solution. Explain the main difference between the purging procedure of a system filled with water and the purging procedure of a system filled with a glycol solution.

_____

_____

_____

118. On a service call, you notice that every time a particular customer's boiler cycles on, water leaks from the pressure-relief valve. What is the most likely reason for this happening? What service should be done to stop it?

_____

_____

_____

119. What steps should be taken if a hydronic system's components are functioning properly, but the system is still providing uneven heating?

_____

_____

_____

Name _____

120. List four signs that indicate that boiler operation is compromised.

_____

_____

_____

11

Notes

Name _____

Date _____ Class _____

> *Carefully study the chapter and then answer the following questions.*

40.1 Heat Pump Basics

_____ 1. *True or False?* Since a heat pump's heat exchanger coils can switch functions, neither coil is permanently called an evaporator or a condenser.

_____ 2. *True or False?* When a heat pump is operating in heating mode, the indoor coil functions as an evaporator.

_____ 3. *True or False?* When a heat pump is operating in heating mode, the outdoor coil functions as a condenser.

_____ 4. *True or False?* In a heat pump system, the direction of refrigerant flow through the compressor remains the same when the system switches between heating and cooling modes.

11

40.2 Types of Heat Pumps

> *Match each of the following types of heat pumps and heat pump components with the phrase that best describes it.*

_____ 5. Heat pump that uses the outside air as a heat source or a heat sink.

_____ 6. Refrigerant coil located inside that warms or cools air in the conditioned space.

_____ 7. Refrigerant coil that transfers heat to and from ambient air.

_____ 8. Type of air-source heat pump that transfers heat between outside air and air inside using forced air for distribution.

_____ 9. Type of air-source heat pump that transfers heat between outside air and air inside using a hydronic distribution system.

_____ 10. Heat pump that uses the earth as a heat source or a heat sink for producing the desired temperature.

_____ 11. Type of heat pump that circulates refrigerant through an outdoor coil placed in direct contact with the earth.

_____ 12. Type of heat pump that circulates water through a secondary circuit that is placed underground or underwater.

_____ 13. Heat pump system that pumps water from and discharges water into a well or lake.

_____ 14. Heat exchanger where heat is transferred between refrigerant from the indoor coil and water from the ground or water loop.

A. Air-source heat pump

B. Air-to-air heat pump

C. Air-to-water heat pump

D. Direct-exchange (DX) heat pump

E. Ground-source heat pump

F. Indoor coil

G. Open-loop heat pump system

H. Outdoor coil

I. Water coil

J. Water-source heat pump

_____ 15. *True or False?* Air-source heat pumps are the most efficient type of heat pump because they access a renewable energy source and have the highest cooling efficiency and lowest annual operating costs.

_____ 16. *True or False?* A well or lake serves as both the supply and discharge source for an open-loop heat pump system.

40.3 Heat Pump Efficiency

_____ 17. *True or False?* The four methods for gauging heat pump efficiency are energy efficiency ratio (EER), seasonal energy efficiency ratio (SEER), heating seasonal performance factor (HSPF), and coefficient of performance (COP).

_____ 18. *True or False?* A 12-SEER heat pump is more efficient than a 13-SEER heat pump.

Name _____

40.4 Heat Pump System Components

Identify each of the principal components of the heat pump system shown in the following illustration. Not all letters will be used.

_____ 19. Accumulator

_____ 20. Compressor

_____ 21. Indoor blower

_____ 22. Indoor coil

_____ 23. Metering device

_____ 24. Outdoor coil

_____ 25. Outdoor fan

_____ 26. Reversing valve

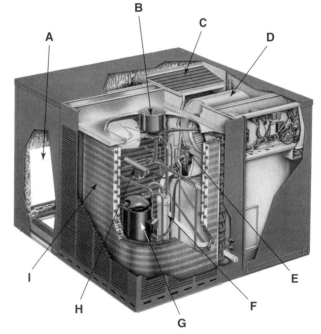

Rheem Manufacturing Company

_____ 27. During the heating mode, some heat pumps store extra liquid refrigerant in a storage device called a(n) _____.
- A. charge compensator tank
- B. liquid storage drum
- C. compressor crankcase
- D. liquid line tank

_____ 28. A heat pump's compressor is protected from liquid slugging by a(n) _____.
- A. liquid line filter-drier
- B. sight glass
- C. slug suppressor
- D. suction line accumulator

_____ 29. A heat pump's crankcase heater warms the compressor to vaporize any liquid refrigerant that may have entered the compressor during _____.
- A. installation
- B. normal cooling mode operation
- C. normal heating mode operation
- D. prior to start-up (On cycle)

11

Match each numbered entry with the term that best describes it.

_____ 30. Meters refrigerant passing through it in one direction and allows refrigerant to pass freely in the other direction.

_____ 31. A fixed orifice metering device that is not well suited for changing conditions.

_____ 32. Uses the pressure difference between the compressor's suction and discharge lines to move its piston.

_____ 33. An arrangement in which a de-energized reversing valve directs high-pressure refrigerant to the indoor coil.

_____ 34. A biflow TXV that allows refrigerant to flow in either direction by using an internal check valve.

_____ 35. A biflow TXV that meters refrigerant flowing in either direction.

_____ 36. An arrangement in which a de-energized reversing valve directs high-pressure refrigerant to the outdoor coil.

A. Biflow bypass TXV

B. Biflow metering TXV

C. Capillary tube

D. Failing into cooling mode

E. Failing into heating mode

F. Flow check piston

G. Pilot-operated reversing valve

Match each numbered entry with the heat pump component that best describes it.

_____ 37. A length of tubing that contacts the ground and circulates water.

_____ 38. A length of tubing that contacts air and circulates refrigerant.

_____ 39. A length of tubing that contacts water and circulates water.

_____ 40. A length of tubing that contacts ground and circulates refrigerant.

_____ 41. A heat exchanger that circulates refrigerant through one passage and water through a separate but adjacent passage.

A. Air coil

B. Ground coil

C. Ground loop

D. Water coil

E. Water loop

Name _____

40.5 Heat Pump Controls

Match each of the following functions with the lettered options used to identify wiring terminals in heat pump thermostats.

_____ 42. Cooling or cooling/heating stage 1   A. O/B

_____ 43. Dehumidifier control   B. OD

_____ 44. Emergency heat or auxiliary heat   C. W1

_____ 45. Fan control   D. DH

_____ 46. Heating stage 1 or auxiliary heat   E. E

_____ 47. Humidity control   F. Y1

_____ 48. Indoor temperature sensor   G. G

_____ 49. Outdoor temperature sensor   H. H

_____ 50. Reverse valve   I. ID

_____ 51. A defrost method that reverses refrigerant flow through an air-source heat pump is called _____ defrost.
   A. auxiliary heat
   B. biflow
   C. reverse cycle
   D. direct-exchange

_____ 52. Why do some air-source heat pumps heat their condensate drain during the defrost cycle?
   A. Government rebate incentives
   B. Increased system energy efficiency
   C. Prevent a frozen drain blockage
   D. Service technician comfort

_____ 53. A heat pump is effective only when the heat pump output equals or exceeds _____.
   A. ambient temperature drop rate
   B. conditioned space heat loss
   C. demand defrost ratio
   D. riser ratio

_____ 54. An air-source heat pump must use auxiliary heat during the following occasions, *except* _____.
   A. during reverse cycle defrost
   B. during emergency heat operation when a compressor is inoperable
   C. when outside temperature is so low that the heat pump cannot transfer enough heat
   D. when a solenoid burns out on a reversing valve arranged as failing into heating mode

11

40.6 Heat Pumps and Solar Heating Systems

_____ 55. *True or False?* Solar panels can be used with a heat pump system to provide auxiliary heat but cannot be used to help heat water for domestic use.

_____ 56. *True or False?* Solar panels cannot be used with a heat pump system to heat water for domestic use.

40.7 Heat Pump System Service

_____ 57. *True or False?* Compared to combustion heating systems, the temperature of the heated air provided by heat pump systems is generally higher at room registers.

_____ 58. Concerning an air-source heat pump's outdoor coil, the most convenient way to deal with the problem of snow accumulation is to _____.
   A. erect a tent that provides total coverage and blocks airflow
   B. install the outdoor coil in the conditioned space
   C. mount the outdoor coil on risers
   D. plant tall, thick bushes surrounding the coil with a 2″ separation gap

_____ 59. After blowing out a drain line with high-pressure air or nitrogen, flush the drain line with _____.
   A. one part household bleach to four parts water
   B. straight cola
   C. seltzer and lemon
   D. steam or kerosene

_____ 60. Ideally, the refrigerant charge of a heat pump should be checked when the outdoor temperature is _____.
   A. above 40°F (4°C) for cooling and below 70°F (21°C) for heating
   B. above 70°F (21°C) for cooling and below 40°F (4°C) for heating
   C. below 40°F (4°C) for cooling and above 70°F (21°C) for heating
   D. below 70°F (21°C) for cooling and above 40°F (4°C) for heating

_____ 61. *True or False?* Running a heat pump on each of its cycles is a good way to check the operation of the reversing valve.

_____ 62. The best way to eliminate bacteria and mold growth in a heat pump system is to _____.
   A. charge the system with the exact amount of refrigerant required
   B. install an ultraviolet light
   C. replace the reversing valve
   D. set the indoor blower to always run on high speed

Name _____

Critical Thinking

63. Explain how a ground-source heat pump is able to produce consistent heating and cooling temperatures throughout the year, despite drastic differences in seasonal weather conditions.

_____

_____

_____

_____

_____

64. Explain why it might be necessary to use a solution of water and glycol in a water-source heat pump.

_____

_____

_____

65. Explain why an air-source heat pump in heating mode operates less efficiently the lower ambient temperature drops.

_____

_____

_____

66. Explain why water loops are submerged at least six feet deep in ponds or lakes.

_____

_____

_____

67. Explain why ice forms on the outdoor coil of an air-source heat pump during operation in cold weather.

_____

_____

_____

_____

68. Explain why a forced-air heat pump's indoor coil is susceptible to mold growth. Explain how mold growth can be prevented.

_____

_____

_____

_____

_____

11

Notes

CHAPTER 41

Gas-Fired Heating Systems

Name _____

Date _____ Class _____

Carefully study the chapter and then answer the following questions.

41.1 Gas Furnace Operation Overview

_____ 1. A gas furnace produces heat for distribution through _____.

    A. combustion

    B. convection

    C. electric sparks

    D. None of the above.

_____ 2. The gas supply to a gas furnace is maintained at a low pressure by the _____.

    A. air break

    B. manifold

    C. pressure regulator

    D. solenoid valve

_____ 3. In some gas furnaces, a device consisting of an inverted opening in the flue that can operate to help prevent back pressure is known as the _____.

    A. air break

    B. manifold

    C. pressure regulator

    D. solenoid valve

41.2 Combustion

_____ 4. A gas furnace that does not burn natural gas is most likely adjusted to burn _____.

    A. crude oil

    B. gasoline

    C. hydrogen

    D. liquified petroleum (LP) gas

_____ 5. Another term for perfect combustion is _____ combustion.

    A. complete

    B. incomplete

    C. integrated

    D. stoichiometric

11

_____ 6. If a gas furnace has a fuel-air mixture with natural gas below 4%, the mixture is considered _____ for combustion.
   A. lean
   B. rectified
   C. rich
   D. soft

_____ 7. If a gas furnace has a fuel-air mixture with natural gas over 14%, the mixture is considered _____ for combustion.
   A. lean
   B. rectified
   C. rich
   D. soft

_____ 8. The important element that must be present in proper amounts and is consumed as it supports combustion is _____.
   A. carbon dioxide
   B. carbon monoxide
   C. nitrogen
   D. oxygen

_____ 9. Air mixed with fuel prior to ignition is _____.
   A. atmospheric
   B. flue air
   C. primary air
   D. secondary air

_____ 10. Air that is added to a flame after ignition is _____.
   A. atmospheric
   B. flue air
   C. primary air
   D. secondary air

_____ 11. Proper combustion in a gas furnace is shown in a flame that burns the color _____.
   A. blue
   B. green
   C. red
   D. white

_____ 12. _True or False?_ Natural gas is lighter than air.

_____ 13. _True or False?_ LP gas produces more heat per pound than natural gas.

_____ 14. _True or False?_ Natural gas requires more oxygen than LP gas for complete combustion.

_____ 15. Many combustion analyzers can be used to measure _____.
   A. carbon dioxide content
   B. carbon monoxide content
   C. stack temperature
   D. All of the above.

Name _____

_____ 16. The term *ultimate carbon dioxide content* refers to the specific amount of carbon dioxide by volume that is present in flue gas when the _____ to achieve complete combustion.

   A. exact amount of air is supplied
   B. maximum amount of excess air is supplied
   C. minimum amount of excess air is supplied
   D. None of the above.

_____ 17. *True or False?* Too much excess air increases the flame temperature in a furnace.

_____ 18. Increasing a furnace's gas pressure can produce the following effects, *except* _____.

   A. a decrease of the flame temperature
   B. a decrease in the oxygen content in the flue gas
   C. incomplete combustion
   D. similar conditions as a lack of air

_____ 19. Furnace gas pressure is typically measured in _____.

   A. in. WC
   B. microns
   C. psi
   D. None of the above.

_____ 20. As the amount of excess air increases, _____.

   A. a furnace's efficiency increases
   B. a furnace's stack temperature decreases
   C. more heat is being transferred in the heat exchanger
   D. more heat is carried through the flue to the outdoors

_____ 21. The stack temperature in a noncondensing gas furnace should be _____.

   A. under 140°F (60°C)
   B. between 140°F and 230°F (60°C and 110°C)
   C. between 230°F and 325°F (110°C and 163°C)
   D. between 325°F and 500°F (163°C and 260°C)

_____ 22. The stack temperature in a condensing gas furnace should be _____.

   A. under 140°F (60°C)
   B. between 140°F and 230°F (60°C and 110°C)
   C. between 230°F and 325°F (110°C and 163°C)
   D. between 325°F and 500°F (163°C and 260°C)

11

41.3 Gas Valves

_____ 23. A single unit that contains multiple components, such as a manual shutoff, adjustments, and a pressure regulator, is called a(n) _____ gas valve.

   A. atmospheric
   B. combination
   C. interlocked
   D. manifold

_____ 24. During normal system operation, a gas valve allows flow primarily when _____.

    A. opened manually

    B. the pilot light goes out

    C. the thermostat calls for heat

    D. None of the above.

_____ 25. *True or False?* LP gas furnaces often have a pressure regulator that is separate from the gas valve.

_____ 26. A gas valve provides low-pressure gas to the _____.

    A. flue

    B. manifold

    C. regulator

    D. shutoff valve

41.4 Gas Burners

_____ 27. Specially sized orifices attached to the gas manifold are _____.

    A. ribbon burners

    B. slotted burners

    C. spuds

    D. None of the above.

_____ 28. *True or False?* At equal pressure, less natural gas will flow through an orifice than LP gas.

_____ 29. A burner that produces a solid flame along the top length of its long, horizontal orifice is a(n) _____ burner.

    A. inshot

    B. ribbon

    C. slotted

    D. None of the above.

_____ 30. A burner that produces a series of individual flames along a row of horizontal orifices is a(n) _____ burner.

    A. inshot

    B. ribbon

    C. slotted

    D. None of the above.

_____ 31. A single-port burner that directs the mixture of air and fuel gas through a large orifice to produce a large flame is a(n) _____ burner.

    A. inshot

    B. ribbon

    C. slotted

    D. None of the above.

Name _____

Match each of the components of the power burner with the proper term.

_____ 32. Air-pressure switch

_____ 33. Blower motor

_____ 34. Gas valve

_____ 35. Ignition control module

_____ 36. Primary air shutter

Midco International, Inc.

41.5 Ignition Systems

_____ 37. The purpose of a standing pilot is to _____.
  A. conduct a high-voltage signal
  B. ignite the gas burner flame
  C. preheat the plenum
  D. None of the above.

_____ 38. *True or False?* If the pilot in a natural gas furnace fails to light, the gas will collect in a low area and create pockets of highly explosive gas.

_____ 39. The safety device used in a standing pilot system to sense the pilot is a _____.
  A. cad cell
  B. sensing bulb and diaphragm switch
  C. thermistor
  D. thermocouple

_____ 40. An intermittent-pilot ignition system burns and monitors its pilot light _____.
  A. all the time
  B. once every 15 minutes during cold weather
  C. only while the thermostat is calling for heat
  D. None of the above.

11

_____ 41. Flame rectification involves using the pilot flame to _____.

    A. change the frequency of a small electric signal
    B. change a small electric current from ac to dc
    C. change a small electric current from dc to ac
    D. limit the current of an electric signal

_____ 42. Flame rectification is used for the purpose of _____.

    A. ionizing primary and secondary air for better combustion
    B. preheating the plenum
    C. providing power for the electronics in the furnace controls
    D. verifying the presence of a pilot flame

_____ 43. When a direct-spark ignition system's electrode and flame rod are packaged together, they provide what is known as _____ sensing.

    A. local
    B. remote
    C. thermal
    D. None of the above.

_____ 44. During normal operation, a disadvantage that only a direct-spark ignition system can produce is _____.

    A. electromagnetic interference (EMI)
    B. a fire hazard
    C. a gas leak
    D. None of the above.

_____ 45. Which gas furnace ignition device is very fragile, sensitive to vibration, and easily broken?

    A. flint striker
    B. hot-surface igniter (glow coil)
    C. indirect pilot lighter
    D. spark igniter

_____ 46. A hot-surface igniter reaches flame ignition temperature through _____.

    A. conduction of electrical current
    B. heat of high-voltage sparks
    C. physical friction
    D. pilot light pre-heating

41.6 Gas Furnace Controls

_____ 47. A gas furnace's ignition control module monitors several _____, which prevent the furnace from operating unless certain conditions are met.

    A. electric interlocks
    B. mechanical linkages
    C. pressure bellows
    D. None of the above.

_____ 48. On modern gas furnaces, the first device that begins operation on a call for heat is the _____.

    A. blower
    B. burner
    C. combustion blower
    D. gas valve

Name _____

_____ 49. A primary difference between nonintegrated ignition controls and integrated ignition controls is the ability to _____.

    A. control ignition
    B. open and close the gas valve
    C. perform self-diagnostic function
    D. None of the above.

_____ 50. An ignition control module has _____ if it closes the gas valve but not the pilot valve when the flame rod does not detect a flame.

    A. 100% shutoff
    B. 100% shutoff with continuous retry
    C. non-100% shutoff
    D. None of the above.

Match each of the furnace component wire terminal designations with its proper term.

_____ 51. Combined gas valve and pilot valve   A. 24 V

_____ 52. Combustion blower   B. 24 V GND

_____ 53. Flame sensor   C. FS or SENSE

_____ 54. Hot-surface igniter   D. GV, MGV, or V1

_____ 55. Main gas valve   E. IGN

_____ 56. Pilot valve   F. IND

_____ 57. Pressure switch   G. MV/PV

_____ 58. Spark igniter   H. PSW

_____ 59. Switched leg of the secondary transformer   I. PV

_____ 60. Unswitched leg of the secondary transformer   J. SPARK

_____ 61. The sheet metal chamber where heat collects before being distributed is called the _____.

    A. blower compartment
    B. bonnet
    C. draft chamber
    D. flue

_____ 62. Flame rollout is when a furnace's flame _____.

    A. reaches through a crack in the heat exchanger and into the plenum
    B. spills backward out of the burner
    C. stretches into the flue
    D. None of the above.

_____ 63. The primary action performed by a rollout switch when it is actuated is to _____.

    A. close the gas valve
    B. stop the blower
    C. stop the combustion blower
    D. stop ignition

11

_____ 64. Which two functions does a combustion blower perform?

    A. It circulates products of combustion and draws in return air.

    B. It draws in conditioned air and circulates return air.

    C. It draws in return air and circulates conditioned air.

    D. It pre-purges the heat exchanger and exhausts flue gases.

_____ 65. A flow switch that uses a large paddle to catch a draft to open or close its contacts is a(n) _____ switch.

    A. end

    B. float

    C. pressure

    D. sail

_____ 66. In a gas furnace application, a switch that connects to the end of a damper motor shaft is a(n) _____ switch.

    A. end

    B. float

    C. pressure

    D. sail

_____ 67. In a gas furnace application, a switch that uses a diaphragm to connect its contacts is a(n) _____ switch.

    A. end

    B. float

    C. pressure

    D. sail

41.7 Gas Furnace Efficiency

_____ 68. The higher the efficiency of a gas furnace, the more likely it is to develop _____.

    A. condensation

    B. electrical shorts

    C. gas leaks

    D. None of the above.

_____ 69. Combustion blowers are used in _____.

    A. all gas furnaces ever

    B. high-efficiency furnaces only

    C. low-efficiency furnaces only

    D. mid-efficiency and high-efficiency furnaces

_____ 70. In low-efficiency gas furnaces, flue gases are moved outdoors by _____.

    A. the combustion blower

    B. the indoor blower

    C. natural convection

    D. the pressure switch

Name _____

> *Match each of the components of this furnace with the proper term.*

_____ 71. Air filter

_____ 72. Burners

_____ 73. Combustion blower

_____ 74. Condensate drain connection

_____ 75. Gas valve

_____ 76. Indoor blower

_____ 77. Pressure switch

_____ 78. Primary heat exchanger

_____ 79. Secondary heat exchanger

_____ 80. Vent outlet

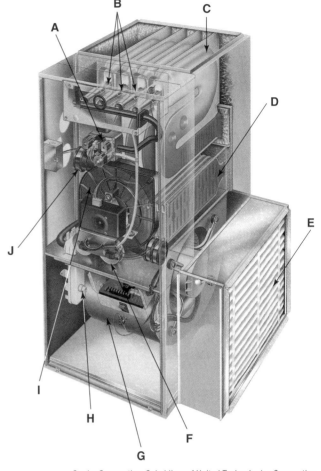

Carrier Corporation, Subsidiary of United Technologies Corporation

41.8 Gas Furnace Venting Categories

_____ 81. The two primary variables that determine a furnace's venting classification are the _____ of the flue gas.

    A. chemical content and pressure

    B. chemical content and temperature

    C. pressure and temperature

    D. pressure and volume

_____ 82. *True or False?* Category I furnaces include low-efficiency and mid-efficiency furnaces.

_____ 83. Category IV furnaces have a _____.

    A. flue gas temperature less than 140°F (60°C)

    B. low concentration of chemicals

    C. low volume flue gas

    D. negative pressure

_____ 84. Category III furnaces require venting made of ____.

    A. galvanized steel

    B. masonry

    C. plastic (PVC or CPVC)

    D. stainless steel

41.9 Gas-Fired Radiant Heat

_____ 85. Radiant gas heaters typically use elements made of ____.

    A. ceramic

    B. lava rock

    C. nichrome

    D. tungsten

_____ 86. In gas-fired radiant heaters, about ____% of the heat energy is converted into radiant heat.

    A. 25

    B. 50

    C. 75

    D. 98

_____ 87. Gas radiant heating is primarily used for ____.

    A. central radiant heating

    B. forced-air heating

    C. hydronic heating

    D. spot heating

41.10 Gas-Fired Heating System Service

_____ 88. Factors that affect pipe size in a gas piping installation include the following, *except* ____.

    A. the amount of gas consumed by the furnace per hour

    B. pressure drop of the piping

    C. piping material

    D. the specific gravity of the gas

_____ 89. A gas piping installation to a furnace or boiler should include a(n) ____ to trap and collect possible contaminants that may flow with the gas.

    A. drip leg

    B. filter-drier

    C. P trap

    D. U bend

_____ 90. After an annual gas furnace inspection, the customer should be reminded to shut off the system and notify the technician if there are ____.

    A. unusual odors in the conditioned space

    B. new or unfamiliar sounds while the furnace is operating

    C. visibly burned components on the furnace

    D. All of the above.

Name _____

_____ 91. On a gas furnace service call when measuring gas pressure or duct static pressures, use a _____.

    A. digital manometer
    B. multimeter
    C. small inspection mirror
    D. thermometer

_____ 92. On a gas furnace service call when measuring flame sensor rectification, use a _____.

    A. digital manometer
    B. multimeter
    C. small inspection mirror
    D. thermometer

_____ 93. If the gas burners are operating when the combustion blower is not operating, the _____ may be ruined.

    A. indoor filter
    B. heat exchanger
    C. flue pipe
    D. pressure switch

_____ 94. A _____ flame indicates that gas velocity is faster than the speed the gas can burn.

    A. lifting
    B. popping
    C. rollout
    D. yellow

_____ 95. A chimney or flue pipe should extend at least _____ above the highest part of a roof.

    A. 1/2″ (1.27 cm)
    B. 2″ (5.08 cm)
    C. 2′ (60.96 cm)
    D. 2 yards (182.88 cm)

_____ 96. The type of venting that is composed of masonry is _____.

    A. Class A
    B. Class B
    C. Class C
    D. None of the above.

_____ 97. The type of venting that is composed of double-wall metal is _____.

    A. Class A
    B. Class B
    C. Class C
    D. None of the above.

_____ 98. *True or False?* High-efficiency furnace flue gases can be vented to the outside through a PVC pipe.

_____ 99. Thermostatic draft regulators open when a bimetal element senses and reacts to the heat of _____.

    A. conditioned air
    B. flue gas
    C. outdoors
    D. None of the above.

Critical Thinking

100. Explain why gas furnaces are designed to take in excess air.

_____

_____

_____

_____

101. Explain how atmospheric gas burners induce airflow through a burner without the use of a blower.

_____

_____

_____

_____

102. Explain the venturi effect and how it is used in gas burners.

_____

_____

_____

_____

103. Explain some of the reasons why standing pilot ignition systems are no longer manufactured and not in use as much as previously.

_____

_____

_____

_____

104. Explain the difference between soft lockout and hard lockout.

_____

_____

_____

_____

105. Briefly describe the differences in venting between mid-efficiency gas furnaces and high-efficiency gas furnaces.

_____

_____

_____

_____

_____

Name _____

106. In your own words, explain the direct-venting system used on a gas furnace.

_____

_____

_____

_____

_____

_____

107. What should a technician look for when visually inspecting a heat exchanger?

_____

_____

_____

_____

108. Explain what problems may be indicated if a furnace's flame characteristics change when the indoor blower turns on.

_____

_____

_____

109. List at least four factors a technician must consider when installing venting for a gas furnace or boiler.

_____

_____

_____

_____

11

Notes

CHAPTER 42

Oil-Fired Heating Systems

Name _____

Date _____ Class _____

Carefully study the chapter and then answer the following questions.

42.1 Basic Oil Furnace Operation

_____ 1. An oil furnace's _____ sends high-voltage electricity to the electrodes.

A. fuel unit
B. igniter
C. stack thermostat
D. system thermostat

_____ 2. An oil furnace's _____ calls for heat to initiate the On cycle.

A. fuel unit
B. igniter
C. stack thermostat
D. system thermostat

_____ 3. An oil furnace's _____ provides combustion air for the furnace.

A. automatic draft regulator
B. blower motor
C. burner motor
D. flue

_____ 4. An oil furnace's _____ turns the pump in the fuel oil.

A. automatic draft regulator
B. blower motor
C. burner motor
D. flue

_____ 5. The operation of the majority of the devices in an oil furnace is initiated by the _____.

A. high-limit control
B. ignition control
C. primary control unit
D. stack thermostat

11

Match the components of the oil furnace shown in the following illustration with the proper term.

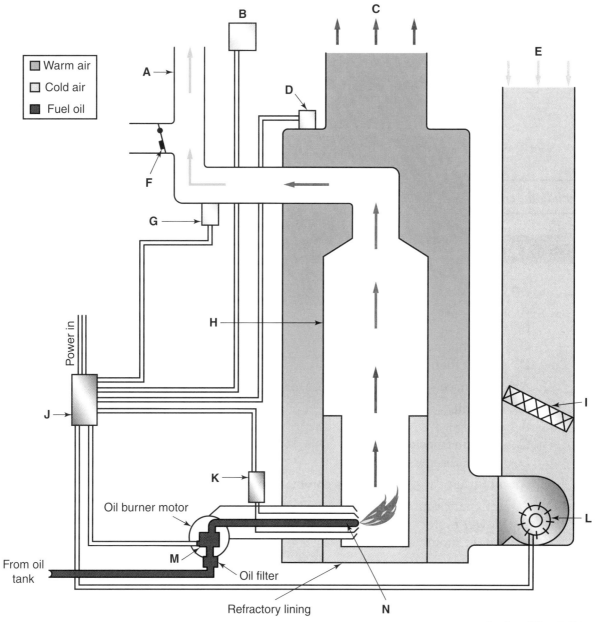

Goodheart-Willcox Publisher

_____ 6. Air filter

_____ 7. Automatic draft regulator

_____ 8. Blower

_____ 9. Burner nozzle

_____ 10. Flue

_____ 11. Fuel unit

_____ 12. Heat exchanger

_____ 13. High-limit control

_____ 14. Ignition control

_____ 15. Primary control unit

_____ 16. Return airflow
(from conditioned space)

_____ 17. Stack thermostat

_____ 18. Supply airflow
(to conditioned space)

_____ 19. System thermostat

Name _____

42.2 Fuel Oil

_____ 20. The type of fuel oil that is most commonly used in residential and light commercial heating applications is ____.

A. No. 1
B. No. 2
C. No. 3
D. No. 4

_____ 21. A fuel oil's ability to vaporize is called its ____.

A. distillation quality
B. flash point
C. pour point
D. viscosity

_____ 22. A liquid's resistance to flow is known as its ____.

A. distillation quality
B. pour point
C. sediment content
D. viscosity

_____ 23. As the temperature of fuel oil decreases, it ____.

A. does not flow more easily
B. exhibits no change in its ability to flow
C. flows more easily
D. None of the above.

_____ 24. The amount of noncombustible contaminants in fuel oil is referred to as ____.

A. ash content
B. carbon residue
C. distillation quality
D. water content

_____ 25. The maximum temperature at which a fuel oil can be stored, retrieved, and used is its ____.

A. distillation quality
B. flash point
C. heating value
D. pour point

11

42.3 Combustion Efficiency

_____ 26. Complete combustion in an oil-fired system is indicated by a flame that is ____.

A. all blue
B. deep red with hints of orange
C. mainly yellow
D. white

_____ 27. For combustion analysis, a low CO_2 reading indicates _____.

    A. complete combustion
    B. incomplete combustion
    C. overfiring
    D. None of the above.

_____ 28. For combustion analysis, a high CO_2 reading indicates _____.

    A. average operation
    B. overfiring
    C. underfiring
    D. None of the above.

_____ 29. The average range of CO_2 content indicating proper combustion in an oil furnace is _____.

    A. 8% to 12%
    B. 12% to 25%
    C. 25% to 38%
    D. 39% to 53%

_____ 30. For accurate readings for a CO_2 test, the analyzer probe should be placed _____.

    A. in the combustion chamber
    B. in the fresh air inlet
    C. near the coldest part of the flue
    D. near the hottest part of the flue

_____ 31. The most common reason why an oil-fired furnace is overfiring is because the _____.

    A. oil pressure setting was decreased
    B. oil pressure setting was increased
    C. original nozzle was replaced with a higher-capacity nozzle
    D. original nozzle was replaced with a lower-capacity nozzle

_____ 32. Stack temperature is the temperature of the flue gas in the flue pipe minus the temperature of the _____.

    A. combustion air
    B. fuel oil
    C. heat exchanger
    D. return air

_____ 33. A high stack temperature measurement could mean the following, *except* _____.

    A. the burner is overfiring
    B. the heat exchanger may not be transferring heat efficiently
    C. there is too little draft
    D. there is too much draft

_____ 34. In an oil-fired furnace, *draft* is described as the movement of _____.

    A. flue gas
    B. fuel oil into the combustion chamber
    C. return and supply air
    D. None of the above.

Name _____

_____ 35. A device installed in the flue pipe that controls the amount of air drawn into the flue is called a(n) _____.
   A. burner blower
   B. draft regulator
   C. high-limit control
   D. stack thermostat

_____ 36. If the reading on a draft gauge is high, the weight on the draft regulator should be moved to _____.
   A. increase draft
   B. reduce draft
   C. stop all draft
   D. None of the above.

_____ 37. Measuring draft in the combustion chamber is part of a(n) _____ draft test.
   A. blower
   B. flue
   C. overfire
   D. stack

_____ 38. Incomplete combustion causes fuel oil that is not fully consumed to turn into _____.
   A. carbon dioxide and carbon monoxide
   B. carbon dioxide and water vapor
   C. smoke and soot
   D. None of the above.

42.4 Fuel Line Components

_____ 39. *True or False?* One-pipe fuel delivery systems have air problems less often than two-pipe systems.

_____ 40. The major benefit of a two-pipe fuel delivery system over a one-pipe system is that a two-pipe system _____.
   A. does not have a return line
   B. does not have a supply line
   C. is self-priming
   D. requires no fuel unit

_____ 41. *True or False?* Air or other gases in the fuel line of an oil-fired heating system may result in increased oil pressure.

11

Match the components of the oil deaerator and fuel unit shown in the following illustration with the proper term.

_____ 42. Bypass plug

_____ 43. Fuel line filter

_____ 44. Fuel oil tank

_____ 45. Fuel unit

_____ 46. Inlet port

_____ 47. Oil deaerator

_____ 48. Outlet to oil burner

_____ 49. Return port

_____ 50. Shutoff valve

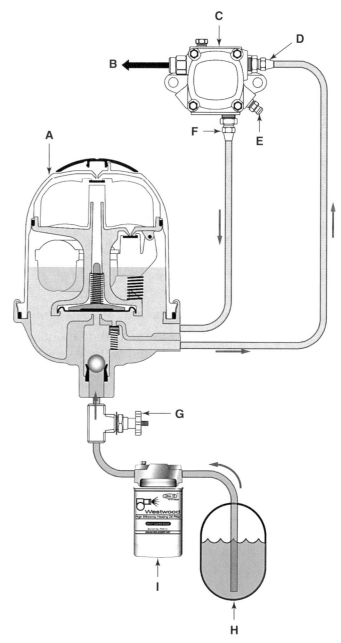

Westwood Products, Inc.

_____ 51. The best place for an oil furnace to have a fuel line filter installed is between the _____.

    A. deaerator and fuel unit

    B. fuel tank and deaerator

    C. fuel unit and oil burner

    D. oil burner and the combustion chamber

Name _____

_____ 52. A booster pump should be located close to the _____ so that it functions to push fuel oil to the reservoir tank.
   A. deaerator
   B. flue
   C. fuel oil tank
   D. oil burner

42.5 Oil Burners

_____ 53. The type of oil burner that uses a motor to operate a fan and a pump is a(n) _____ burner.
   A. gun
   B. inshot
   C. pot
   D. ribbon

> *Match the components of the oil burner shown in the following illustration with the proper term.*

_____ 54. Adjustable air band

_____ 55. Air tube and mounting flange

_____ 56. Blower wheel

_____ 57. Burner housing

_____ 58. Cad cell

_____ 59. Combustion head

_____ 60. Coupling

_____ 61. Electrodes

_____ 62. Fuel line

_____ 63. Fuel unit (oil pump)

_____ 64. Igniter

_____ 65. Motor and capacitor

_____ 66. Primary control unit

_____ 67. Wiring box

Carlin Combustion Technology, Inc.

11

_____ 68. A static pressure disk is used to _____.

   A. control fuel oil flow
   B. disturb airflow and create air turbulence
   C. modulate system draft
   D. All of the above.

_____ 69. A combustion head is a plate with slots and holes that are designed to promote ideal combustion by _____.

   A. directing airflow into the combustion chamber
   B. guiding return air around the heat exchanger
   C. increasing combustion chamber pressure
   D. regulating draft

_____ 70. *True or False?* A flame retention oil burner produces a cleaner and hotter flame than a standard oil burner.

_____ 71. When installing electrodes on an oil burner, be sure that the _____.

   A. electrodes are thoroughly lubricated
   B. electrode tips are in contact with each other
   C. gaps are properly set according to manufacturer specifications
   D. mounting allows the electrodes to rotate freely

_____ 72. To prevent unintentional arcing across electrode insulators, be certain to _____.

   A. clean off any soot from the electrode insulators
   B. encase the electrode insulators in a thick foam wrap
   C. replace dirty insulators with uninsulated electrodes
   D. thoroughly cover the electrode insulators with a film of soot

_____ 73. The process of breaking up fuel oil into tiny droplets is known as _____.

   A. atomization
   B. combustion
   C. evaporation
   D. fractionation

_____ 74. The amount of fuel delivered to the combustion chamber is measured in gallons per hour (gph) at _____.

   A. 10 in. WC
   B. 10 psi
   C. 100 kPa
   D. 100 psi

_____ 75. A burner nozzle that produces little or no droplets in the center is a _____ cone nozzle.

   A. dual
   B. hollow
   C. semisolid
   D. solid

_____ 76. Oil burners used in residential applications usually have _____ motors.

   A. ECM or shaded-pole
   B. permanent split capacitor or split-phase
   C. three-phase or VFD-controlled
   D. None of the above.

Name _____

_____ 77. An oil burner fan is primarily used to _____.

    A. circulate return and supply air
    B. dislodge particulates clinging to the inside of the combustion chamber
    C. prevent the burner from overheating
    D. provide combustion air

_____ 78. An adjustable air band on an oil burner is used to regulate the amount of air that is drawn in and blown through the _____.

    A. air tube
    B. conditioned space
    C. fuel line
    D. None of the above.

_____ 79. An oil burner's fuel unit performs the following tasks, *except* _____.

    A. act as a secondary filter after the fuel line filter
    B. move fuel oil from the fuel oil tank to the oil burner
    C. provide turning torque for the oil burner fan
    D. regulate the pressure of the fuel oil pumped to the oil burner

_____ 80. *True or False?* Two-stage pumps create greater suction than single-stage pumps.

_____ 81. A fuel unit uses a(n) _____ to quickly shut off the flow of fuel.

    A. fast-acting bypass spring
    B. float-operated needle valve
    C. reversing function
    D. solenoid valve

42.6 Primary Control Units

_____ 82. A primary control unit initiates oil furnace operation when a _____ calls for heat.

    A. customer
    B. high-limit control
    C. stack thermostat
    D. system thermostat

_____ 83. A cad cell confirms that an oil furnace has a flame by sensing its _____.

    A. flue gas chemical composition
    B. heat
    C. light
    D. pressure

_____ 84. A stack relay confirms that an oil furnace has a flame by sensing its _____.

    A. flue gas chemical composition
    B. heat
    C. light
    D. pressure

11

_____ 85. A cad cell is primarily installed in an oil burner's _____.

   A. air tube
   B. burner nozzle
   C. fuel unit
   D. primary control

_____ 86. A stack relay is installed in the _____.

   A. combustion chamber
   B. flue
   C. fresh air inlet
   D. return air ductwork

_____ 87. A cad cell reacts to an oil furnace flame by _____.

   A. dropping its resistance
   B. illuminating
   C. producing a voltage
   D. vibrating at a set frequency

Match the components of the ignition transformer with the proper term.

_____ 88. Arc

_____ 89. Coils

_____ 90. Electrode insulator

_____ 91. Electrode rod

_____ 92. Line voltage

_____ 93. Spring terminal

_____ 94. Terminal insulator

_____ 95. Transformer core

NDRA

_____ 96. A solid-state igniter uses _____ to create a spark for oil furnace ignition.

   A. electronic circuitry
   B. friction
   C. mechanical stress on a crystal
   D. a step-up transformer

_____ 97. *True or False?* The higher the voltage produced by an igniter, the higher the temperature of the electric arc.

Name _____

_____ 98. The time when an oil furnace is awaiting a call for heat is referred to as ____.

  A. preignition
  B. recycle time
  C. standby
  D. trial for ignition

_____ 99. The length of time, in seconds, that an igniter will continue to spark after a flame has been sensed is called ____.

  A. ignition carryover
  B. preignition
  C. standby
  D. trial for ignition

_____ 100. The amount of time for a primary control unit to sense that there is no flame when a thermostat is calling for heat is called the ____ time.

  A. flame failure response
  B. post-purge
  C. recycle
  D. trial for ignition

_____ 101. The amount of time, in seconds, before an igniter attempts ignition again after a failed ignition attempt is called ____ time.

  A. flame failure response
  B. post-purge
  C. recycle
  D. trial for ignition

_____ 102. The number of times a furnace will try to ignite for a single call for heat before locking out is called ____ limit.

  A. flame failure
  B. recycle
  C. reset
  D. trial for ignition

_____ 103. The number of times a primary control unit can be reset from a lockout is the ____ limit.

  A. flame failure
  B. recycle
  C. reset
  D. trial for ignition

11

42.7 Oil Furnace Exhaust

_____ 104. *True or False?* The oil burner nozzle should be centered halfway between the top and the bottom of the combustion chamber.

_____ 105. The spray pattern and amount of fuel oil blown into an oil furnace's combustion chamber is primarily determined by the ____.

  A. air tube
  B. burner nozzle
  C. flue
  D. primary control

_____ 106. An oil-fired appliance's combustion chamber is lined with a fire-resistant material called _____ material.

  A. combustion
  B. flame retention
  C. recycle
  D. refractory

_____ 107. Oil furnace venting that relies on the heat flow of the gases is referred to as _____ venting.

  A. barometric
  B. forced draft
  C. induced draft
  D. natural convection

42.8 Oil-Fired Heating System Service

_____ 108. According to the IMC, a manual shutoff valve must be installed _____.

  A. between the burner and the combustion chamber
  B. between the fuel oil tank and burner
  C. between the fuel unit and the deaerator
  D. in the return line

_____ 109. The primary reason that a fuel tank should not be more than 25′ (7.5 m) above the oil burner it is supplying is to _____.

  A. prevent fuel line pressure from exceeding 10 psi (70 kPa)
  B. prevent the fuel tank from falling onto the burner
  C. minimize the material and cost of fuel line installation
  D. None of the above.

_____ 110. A fuel oil tank's vent pipe must be designed with a _____ bend to keep out dirt and rain.

  A. 45°
  B. 90°
  C. 180°
  D. 360°

_____ 111. The copper tubing used in residential oil-fired heating systems is usually connected using _____.

  A. arc welding
  B. brazing
  C. flared fittings
  D. solvent welding

_____ 112. The ignition of a large amount of vaporized fuel oil that blows soot backward out of the combustion chamber is called _____.

  A. backfiring
  B. blowback
  C. flue spew
  D. overfiring

Name _____

_____ 113. Oil furnace installations that should have their fuel lines bled annually are _____ fuel delivery systems.

   A. one-pipe
   B. two-pipe
   C. four-pipe
   D. None of the above.

_____ 114. When handling burner nozzles, _____.

   A. always pick them up by the strainer or orifice
   B. do not twist the air tube or move it out of line
   C. use an awl to clean a dirty or plugged nozzle
   D. All of the above.

_____ 115. Delayed ignition and blowback may be caused by the following, *except* _____.

   A. cracked or poor insulation on the electrodes
   B. high voltage across the electrodes
   C. improper gaps between the electrodes and also the gas burner nozzle
   D. a weak or failing ignition transformer

_____ 116. If a fuel line is found to be dirty and clogged, it is best to blow it out with _____.

   A. ammonia
   B. compressed air
   C. nitrogen
   D. oxygen

Critical Thinking

117. Explain three possible causes of low CO_2 measurements in flue gas.

_____

_____

_____

_____

_____

118. Explain two possible causes of high CO_2 measurements in flue gas.

_____

_____

_____

_____

119. Explain what happens to a furnace's draft and overall system operation and efficiency if indoor air pressure is much greater than outdoor air pressure.

_____

_____

_____

11

120. Explain what may happen to an operating oil furnace if indoor air pressure is too low compared to outdoor air pressure.

_____

_____

121. Explain the function of a fuel unit's bypass plug. Explain what happens when it is installed and when it is not installed. When is a bypass plug installed and not installed?

_____

_____

_____

_____

122. Explain what may occur with an oil burner's ignition spark if each electrode's insulator is covered with soot. Explain why this happens.

_____

_____

_____

123. Explain the difference between interrupted and intermittent ignition.

_____

_____

_____

124. Explain what occurs during preignition.

_____

_____

125. List some of the factors that affect oil furnace draft. Explain how these factors affect draft.

_____

_____

_____

_____

126. What is a draft regulator? What is it intended to do? How does it accomplish its purpose? Explain how its actions affect the system.

_____

_____

_____

_____

_____

_____

Name _____

127. Explain why the cap on a storage tank fill pipe should always be in place.

_____

_____

128. Explain how condensation problems in aboveground storage tanks can be alleviated.

_____

_____

_____

_____

129. Explain why the fuel line must be completely purged of air before an oil burner is started.

_____

_____

_____

130. Explain what a technician should do if a puddle of fuel oil is found in a combustion chamber during an inspection.

_____

_____

_____

131. If a technician suspected that air was in a fuel line as a result of a leak, explain how these suspicions might be verified.

_____

_____

_____

_____

132. List three or more reasons for using fuel oil additives.

_____

_____

_____

_____

133. Briefly explain some of the results of sulfuric acid forming in the flue of an oil-fired heating system.

_____

_____

_____

11

Notes

CHAPTER **43**

Electric Heating Systems

> *Carefully study the chapter and then answer the following questions.*

43.1 Principles of Electric Resistance Heating

_____ 1. Electric heating produces heat as electrons _____.

    A. alternate at a reduced frequency

    B. overcome a circuit's inductance

    C. overcome an element's resistance

    D. None of the above.

_____ 2. One watt is equal to _____.

    A. 3.413 Btu/hr

    B. 341.3 Btu/hr

    C. 3,413 Btu/hr

    D. All of the above.

_____ 3. A 10 A electric heater is powered by a 120 V circuit. What is the maximum amount of heat energy that can be produced by the heater? Show your work.

    A. 3.52 Btu/hr

    B. 32.2 Btu/hr

    C. 351.6 Btu/hr

    D. 4,095.6 Btu/hr

11

_____ 4. A single-phase, line-voltage electric heater is rated at 1,800 W. What is its heat output? Show your work.

    A. 527.4 Btu/hr

    B. 1,796.6 Btu/hr

    C. 1,803.4 Btu/hr

    D. 6,143.4 Btu/hr

_____ 5. How much electrical power is needed to heat a room that requires 7,500 Btu/hr? Show your work.

    A. 2.2 W

    B. 25.6 W

    C. 2,197.5 W

    D. 25,597.5 W

43.2 Electric Heating Elements

_____ 6. Open wire heating elements are often formed into _____, allowing more wire to be used in a smaller space.

    A. coils

    B. knots

    C. webs

    D. All of the above.

_____ 7. _True or False?_ Open ribbon heating elements provide more surface area for air contact than uncoiled open wires.

_____ 8. Electric heating elements are often made of _____.

    A. cast iron

    B. manganese cadmium

    C. nickel chromium

    D. tungsten

_____ 9. The type of electric heating element that may be most easily placed in a fin-type aluminum casting for increased surface area is _____.

    A. open ribbon

    B. open wire

    C. tubular cased wire

    D. All of the above.

Name _____

_____ 10. The type of electric heating element that presents the least danger of electric shock is _____.

    A. open ribbon

    B. open wire

    C. tubular cased wire

    D. All of the above.

43.3 Electric Heating Systems

Match each of the components of the electric furnace shown in the following illustration with the proper term.

_____ 11. Air filter

_____ 12. Contactor

_____ 13. Indoor blower

_____ 14. Open wire heating elements

_____ 15. Transformer

_____ 16. Wiring terminal box

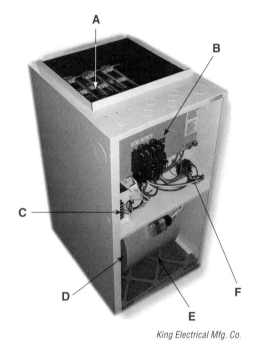

King Electrical Mfg. Co.

_____ 17. Each heating element in an electric furnace consumes _____.

    A. 5 W to 10 W

    B. 5 kW to 10 kW

    C. 17 W to 34 W

    D. 17 kW to 34 kW

_____ 18. An appliance setting at a certain level of heat output is referred to as a _____.

    A. heat sequence

    B. heat stage

    C. heat tier

    D. None of the above.

_____ 19. *True or False?* Electric heating elements are position-sensitive.

_____ 20. *True or False?* Some types of relays used in an electric furnace must be installed in a certain orientation to operate properly.

_____ 21. *True or False?* Most heating elements will not perform at capacity when electrical polarity is reversed.

> *Match each of the components of the duct heater shown in the following illustration with the proper term.*

_____ 22. Fuses

_____ 23. Heating elements

_____ 24. Pressure switch

_____ 25. Magnetic contactor

Tutco, Inc.

_____ 26. *True or False?* Duct heaters can only be used to provide supplementary heat.

_____ 27. Fan heaters primarily distribute heat through _____.

    A. convection
    B. conduction
    C. radiation
    D. All of the above.

_____ 28. *True or False?* A baseboard heating unit rated at 120 V should be connected to a 240 V circuit.

_____ 29. Electric baseboard heating units primarily provide heat through _____.

    A. conduction
    B. forced convection
    C. natural convection
    D. radiation

_____ 30. For baseboard heating units with the same heat-producing capacity, the higher the supplied voltage, the _____.

    A. fewer amps pulled by the circuit
    B. higher the cost of electrical power
    C. lower the cost of electrical power
    D. more amps pulled by the circuit

_____ 31. For direct radiant heat, the heating elements are mounted in a _____.

    A. ceiling
    B. fixture
    C. floor
    D. wall

_____ 32. *True or False?* Direct radiant heat is also called surface radiant heat.

Name _____

_____ 33. Some electric heat and deicing installations are considered continuous load, which requires the use of conductors and overcurrent circuit protection rated at _____ of the continuous load.

   A. 125%
   B. 150%
   C. 200%
   D. 400%

43.4 Electric Furnace and Duct Heater Controls

_____ 34. Before energizing heating elements, controls for furnaces and duct heaters often check for _____.

   A. airflow
   B. applied voltage
   C. electrical continuity
   D. overcurrent protection

_____ 35. A pressure switch may actuate based on a _____.

   A. difference in pressure
   B. drop in pressure
   C. rise in pressure
   D. All of the above.

_____ 36. A sequencer uses a _____ to open and close its sets of contacts.

   A. bimetal element
   B. crystal
   C. diaphragm switch
   D. sail switch

_____ 37. Of the different sets of contacts on sequencers, the contacts that energize the _____ must conduct the most current.

   A. blower fan
   B. heating elements
   C. next sequencer coil
   D. All of the above.

_____ 38. The best way to prevent an electric furnace's indoor blower from turning off immediately once the call for heat has ended is to install a(n) _____.

   A. jumper across any switch controlling the blower
   B. manual blower switch
   C. temperature-sensitive off-delay switch
   D. Any of the above.

_____ 39. A high-temperature safety cutoff is usually a fusible link or a(n) _____.

   A. air pressure switch
   B. bimetal disc
   C. mercury contactor
   D. sail switch

11

_____ 40. After determining why a fusible link tripped off a system, a technician should _____.

    A. install a new replacement fusible link

    B. press the fusible link's reset button

    C. solder the fusible link back together

    D. wire in a jumper around the fusible link

43.5 Electric Baseboard Heating Unit Controls

_____ 41. When an electric baseboard heating unit operates is primarily governed by its _____.

    A. circuit breaker

    B. fusible links

    C. manual switch

    D. thermostat

_____ 42. The majority of residential electric baseboard heating units run on _____.

    A. 12 Vdc

    B. 24 Vac

    C. 120 Vac

    D. 240 Vac

_____ 43. *True or False?* Electric baseboard units commonly share an electrical circuit with several electrical sockets, a room's light fixture, and at least one major appliance (refrigerator, washing machine, etc.).

43.6 Electric Heat Construction Practices

_____ 44. *True or False?* A danger of electric heat is that the highest temperatures reached commonly exceed the ignition temperature of household materials.

_____ 45. *True or False?* Due to the complexity of electric heating systems, its equipment usually requires more physical space than the equipment for other types of heating systems.

_____ 46. *True or False?* Ductwork is necessary for all types of electric heating systems.

_____ 47. Electric heating systems are considered _____ efficient.

    A. 68% to 74%

    B. 78% to 83%

    C. 90% to 98%

    D. 100%

_____ 48. *True or False?* The controlled combustion in electric heating systems results in the production of gases that could accumulate to present toxic conditions indoors.

_____ 49. *True or False?* To minimize heat loss, appropriate insulation should be installed throughout a building that uses electric heat.

_____ 50. In a tightly sealed and insulated building that uses only electric heating, indoor air has a tendency to _____.

    A. drop extremely low in humidity

    B. drop rapidly in temperature

    C. rise in humidity

    D. None of the above.

Name _____

_____ 51. During cold weather, the most energy efficient way to manage temperature and humidity in a tightly sealed and insulated building with electric heat is to ____.

    A. crack open the top half of a Dutch door
    B. install and use an ERV or HRV
    C. leave a window open
    D. use economizer operation to vent stale air and draw in fresh air

43.7 Electric Heating System Service

_____ 52. In a properly installed electric furnace that is de-energized, an ohmmeter should read ____ between each heating element and ground in the control panel.

    A. $0\ \Omega$
    B. $5\ \Omega$ to $15\ \Omega$
    C. $100\ \Omega$ to $250\ \Omega$
    D. OL (infinity)

_____ 53. If a baseboard heating unit's high-limit switch fails and a replacement of the same rating is not available, then ____.

    A. install a limit switch with a 25% higher cut-out setting
    B. install a timer that turns off the unit for 10 minutes every hour
    C. operate the unit without a high-limit switch
    D. replace the entire baseboard unit

_____ 54. The first step to replacing a baseboard heating unit is to turn off the power by ____.

    A. disconnecting the thermostat wiring terminals
    B. opening the circuit breaker
    C. simultaneously cutting both power wires at the unit
    D. None of the above.

_____ 55. If an electric furnace's blower that is driven by a single-phase motor will not run but voltage is applied to it, then mostly likely the ____.

    A. blower's capacitor must be open or shorted
    B. fan coil contacts must be open
    C. thermostat switch must be open
    D. All of the above.

11

Critical Thinking

56. List five advantages of electric heating systems compared to combustion heating systems.

_____

_____

_____

_____

_____

57. In a large warehouse that is mostly storage with two small open areas where employees spend most of their time, explain why it may be more cost effective to use direct radiant heat, rather than forced-air heating.

_____

_____

_____

_____

58. Concerning an electric furnace, some system thermostat switches are wired in series with only one fan coil and one sequencer coil. Other system thermostat switches are wired in series with all the sequencer coils, though the sequencer coils are in parallel with each other. What characteristic would need to be different about the thermostat switch between these two wiring arrangements? Explain how these two wiring arrangements affect system operation.

_____

_____

_____

_____

_____

59. For what reasons are magnetic contactors used less often in electric furnaces and duct heaters than other methods of energizing the heating elements?

_____

_____

_____

60. Explain why it is preferable for a thermostat to use a double-pole switch (instead of a single-pole switch) to control a circuit powering an electric baseboard heating unit.

_____

_____

_____

_____

61. List the relevant information that a technician must check when preparing to replace an electric heating element.

_____

_____

_____

_____

CHAPTER 44

Solar Power and Thermal Storage

Name _____

Date _____ Class _____

Carefully study the chapter and then answer the following questions.

44.1 The Nature of Solar Energy

_____ 1. Solar energy can be described as a stream of ____.

    A. electrons
    B. photons
    C. protons
    D. All of the above.

_____ 2. The shorter the wavelength, the ____ of the wave.

    A. greater the energy
    B. higher the amplitude
    C. lesser the energy
    D. lower the amplitude

_____ 3. Most ultraviolet radiation is absorbed by ____.

    A. bodies of water
    B. green plant material
    C. humans
    D. ozone

_____ 4. Light radiation with wavelengths that are greater than 740 nanometers (nm) is called ____.

    A. infrared radiation
    B. nuclear radiation
    C. ultraviolet radiation
    D. visible light

_____ 5. Light radiation with wavelengths that are less than 380 nanometers (nm) is called ____.

    A. infrared radiation
    B. nuclear radiation
    C. ultraviolet radiation
    D. visible light

12

_____ 6. Light radiation with wavelengths that are between 380 nanometers (nm) and 740 nanometers (nm) is called ____.

   A. infrared radiation
   B. nuclear radiation
   C. ultraviolet radiation
   D. visible light

_____ 7. The solar constant is the energy flow outside the atmosphere of approximately ____.

   A. 1.35 kW/m^2
   B. 3.14 kW/m^2
   C. 3.413 kW/m^2
   D. 200 kW/m^2

_____ 8. As an object gets hotter, the wavelength at which the most radiation is given off becomes ____.

   A. colder
   B. distorted
   C. longer
   D. shorter

44.2 Solar Collectors

_____ 9. A solar energy collector that uses a series of flat-plates with an insulated surface to collect solar energy is a ____.

   A. convex sandwich collector
   B. dished collector
   C. flat-plate collector
   D. tubular collector

_____ 10. A black object's temperature can be increased by placing a(n) ____ cover over its black surface.

   A. brushed steel
   B. glass
   C. mirrored
   D. rubber

_____ 11. A special absorber surface used to increase the temperature of a collector is described as a(n) ____.

   A. adsorbant surface
   B. heating surface
   C. reflective surface
   D. selective surface

_____ 12. A transparent cover used on most flat-plate collectors is expected to fulfill the following functions, *except* ____.

   A. prevent the escape of heat collected by the absorber surface
   B. protect the absorber surface from the weather
   C. transmit sunlight to the absorber surface
   D. None of the above.

Name _____

_____ 13. For all-season home hot water heating, a solar collector should be angled upward about _____ from the horizontal.
  A. 35° to 45°
  B. 67° to 75°
  C. 84° to 90°
  D. 100° to 113°

_____ 14. An efficient type of solar collector that uses tubes containing fluid contained within slightly larger glass tubes that are in vacuum describes _____.
  A. dished vacuum collectors
  B. evacuated tube collectors
  C. flat-plate collectors
  D. tube-within-a-tube collectors

44.3 Solar Heating Systems

_____ 15. *True or False?* In a passive solar energy system, the solar energy is absorbed into a collector.

_____ 16. A solarium or other passive solar heating system relies heavily on floors and walls that have a _____.
  A. high thermal mass
  B. low thermal mass
  C. reflective surfaces
  D. None of the above.

Match each of the components of the Trombe wall with the correct term.

_____ 17. Cool air

_____ 18. Glass

_____ 19. Heat radiated from thermal mass

_____ 20. Natural convection

_____ 21. Trombe wall

_____ 22. Warmed air

Goodheart-Willcox Publisher

12

_____ 23. Passive liquid-based solar heating systems use water's natural tendency to _____.

    A. fall as it heats and rise as it cools
    B. melt as it heats and freeze as it cools
    C. rise as it heats and fall as it cools
    D. None of the above.

_____ 24. A drainback system uses _____ to remove liquid from the collectors whenever the pump is off.

    A. gravity
    B. peristalsis
    C. positive pressure
    D. vacuum

44.4 Applications for Solar Heating Systems

Match each of the components of the two-tank solar water heating system with the correct term.

_____ 25. Cold water in

_____ 26. Conventional water heater

_____ 27. Heat exchanger

_____ 28. Hot water out

_____ 29. Preheated water

_____ 30. Pump

_____ 31. Solar collector

_____ 32. Solar water heater

Goodheart-Willcox Publisher

_____ 33. In solar domestic hot water (DHW) systems, antifreeze _____.

    A. does not come into contact with the water supply
    B. freely mixes with the domestic water
    C. is never used
    D. All of the above.

_____ 34. _True or False?_ In warm climates, the solar heating system for a swimming pool always requires protection from freezing temperatures.

Name _____

44.5 Supplementary Heat

_____ 35. When an air-based solar heating system uses electric resistance heating to supplement heat, _____.
    A. the burner runs only when necessary
    B. a direct electric resistance radiator may be installed in the air ducts
    C. an electric heating element is installed in an auxiliary tank
    D. None of the above.

_____ 36. When a liquid-based solar heating system uses electric resistance heating to supplement heat, _____.
    A. auxiliary burners may be added to heat the water in an auxiliary tank
    B. a direct electric resistance radiator may be installed in the air ducts
    C. an electric heating element can be installed in an auxiliary tank
    D. All of the above.

Match each of the components of the combination forced-air heating system used with a closed liquid-based solar heating system with the correct term.

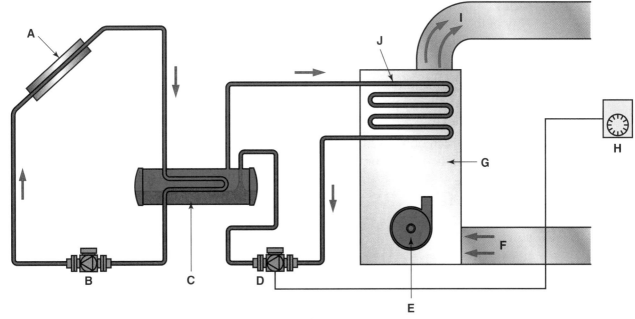

Goodheart-Willcox Publisher

_____ 37. Auxiliary furnace

_____ 38. Blower

_____ 39. Cold air

_____ 40. Hot air

_____ 41. Hot water heat exchanger

_____ 42. Hot water pump

_____ 43. Hot water storage tank

_____ 44. Solar collector

_____ 45. Solar collector pump

_____ 46. Thermostat

_____ 47. *True or False?* If oil or gas is used to supplement heat to a solar heating system, a separate furnace is installed.

44.6 Converting Solar Energy to Electricity

_____ 48. The device most commonly used to convert radiant energy to electricity is a ____.

A. solar cell
B. solar collector
C. solar pump
D. sun generator

_____ 49. Solar modules are constructed into a ____.

A. solar array
B. solar bank
C. solar matrix
D. solar reservoir

_____ 50. *True or False?* A solar array can be used to power a dc electrical device.

_____ 51. Solar cells can be connected in ____ to obtain higher voltages.

A. parallel
B. a positive-to-positive and negative-to-negative terminal arrangement
C. series
D. Any of the above.

_____ 52. A solar energy system that includes batteries in its circuit ____.

A. is ideal for roadside signs in remote areas
B. may be used to help power satellites
C. works well for many portable electronics
D. All of the above.

_____ 53. A program in which homeowners connected to the power grid can sell excess energy they generate back to the utility company is called ____.

A. net metering
B. reverse metering
C. solar sellback
D. sun money

44.7 Solar Energy Cooling Systems

_____ 54. A solar cooling system usually uses ____ collectors.

A. concentrating
B. evacuated tube
C. flat-plate
D. tube-within-a-tube

_____ 55. A solar absorption air-conditioning system can produce more cooling when there is ____.

A. less sun
B. more sun
C. overcast skies
D. a solar eclipse

Name _____

_____ 56. For a solar PV system to be off-the-grid and able to operate standalone, it must have _____.

    A. a bank of batteries and an inverter

    B. a connection to the electrical grid and an inverter

    C. just a connection to the electrical grid

    D. just an inverter

_____ 57. For a solar PV system to engage in net metering, it must have _____.

    A. a bank of batteries and an inverter

    B. a connection to the electrical grid and an inverter

    C. just a connection to the electrical grid

    D. just an inverter

44.8 Thermal Energy Storage (TES) Systems

_____ 58. *True or False?* The term *thermal energy storage* (TES) refers to the temporary storage of heat energy for later use.

_____ 59. The process used by a sensible TES system _____ of the storage medium.

    A. changes the physical phase

    B. mixes an incoming substance with another substance

    C. only changes the pressure

    D. raises and lowers the temperature

_____ 60. The process used by a sensible TES system _____ of the storage medium.

    A. changes the physical phase

    B. mixes an incoming substance with another substance

    C. only changes the pressure

    D. raises and lowers the temperature

_____ 61. A combination of inorganic salts, water, and various elements that are formulated to freeze at a desired temperature is called _____.

    A. crystal water

    B. dry ice

    C. eutectic salts

    D. glycol slurry

_____ 62. A mechanical HVAC system that utilizes ice storage as its CTES system saves money by running its _____ during off-peak hours.

    A. compressor

    B. crankcase heater

    C. indoor blower

    D. All of the above.

_____ 63. A mechanical HVAC system that utilizes ice storage as its CTES system reduces its electrical demand during peak hours by *not* running its _____.

    A. compressor

    B. indoor blower

    C. low-power refrigerant pump

    D. All of the above.

12

Critical Thinking

64. In your own words, briefly explain the process of *electrodeposition*.

_____

_____

_____

65. Explain why a collector surface must be properly insulated.

_____

_____

_____

66. Explain why plastic is considered an inferior collector cover material compared to glass.

_____

_____

_____

67. Briefly explain why the evacuated tube collector is the most efficient type of solar collector.

_____

_____

_____

68. Explain why flat-plate collectors are used more frequently than mirror- or lens-type concentrating collectors.

_____

_____

_____

_____

69. Briefly explain why passive solar energy systems require less energy input than active systems.

_____

_____

_____

_____

Name _____

70. Briefly explain how a Trombe wall functions to heat a conditioned space.

_____

_____

_____

_____

_____

_____

71. Explain the flow of fresh air and natural ventilation in a building that uses both a solar chimney and an earth tube.

_____

_____

_____

_____

_____

_____

72. Explain the reason why most liquid-based solar heating systems use a mixture of water and glycol instead of pure water.

_____

_____

73. Briefly explain how a TES system reduces demand on the electrical grid.

_____

_____

_____

12

Notes

Name _____

Date _____ **Class** _____

Carefully study the chapter and then answer the following questions.

45.1 Energy Consumption

_____ 1. The passing of outside air into a building is referred to as _____.

  A. exfiltration
  B. infiltration
  C. stratification
  D. ventilation

_____ 2. The passing of inside air out of a building is referred to as _____.

  A. exfiltration
  B. infiltration
  C. stratification
  D. ventilation

_____ 3. The forced airflow that takes place by design between the inside and outside of a building is referred to as _____.

  A. exfiltration
  B. infiltration
  C. stratification
  D. ventilation

_____ 4. Occupants in a building can be a source of _____.

  A. latent and sensible heat
  B. latent heat
  C. sensible heat
  D. None of the above.

_____ 5. The heat generated by lighting is in direct proportion to its _____ rating.

  A. lumens
  B. voltage
  C. wattage
  D. None of the above.

12

45.2 Energy Audits

_____ 6. *True or False?* The end result of an energy audit should be the reduction of a building's energy cost and usage.

_____ 7. To test a home for air infiltration and locate leaks, perform a _____.
   A. blower door test
   B. combustion efficiency test
   C. electrical power usage review
   D. insulation measurement

_____ 8. A rough estimate of a home's energy usage can be done by _____.
   A. comparing its energy usage to similar houses in the same locality
   B. examining a year's worth of utility bills
   C. reviewing its year-round thermostat settings
   D. None of the above.

_____ 9. *True or False?* For a residential energy audit, it is important to check the age and maintenance of the heating and cooling system, but you should not bother examining ductwork, registers, and dampers.

_____ 10. *True or False?* An energy audit for a large commercial or industrial complex can be done by a single auditor and take the same amount of time and effort as required for a single-family home residence.

_____ 11. Energy costs for a commercial or industrial building may be dependent on _____.
   A. day of the week
   B. time of day
   C. total amount of energy used at a given time
   D. All of the above.

45.3 Building Control Systems

_____ 12. Total energy management is a conservation concept in which a building is viewed in terms of _____.
   A. the energy used by one subsystem
   B. its total energy usage
   C. the requirements of separate systems
   D. None of the above.

_____ 13. *True or False?* A building control system increases building efficiency by causing each subsystem to operate at peak efficiency without compromising functionality of any of the other subsystems.

_____ 14. *True or False?* An energy audit is concerned with energy usage and does not take into account the weather data where a building is located.

_____ 15. Energy use intensity (EUI) shows a building's annual energy use per _____.
   A. cubic foot
   B. day
   C. square foot
   D. year

Name _____

45.4 Controllers for Building Control Systems

_____ 16. Direct digital control (DDC) is a type of control system that utilizes _____ to control an HVAC or automated building system.
A. analog and digital electric signals
B. hydraulic signals
C. pneumatic signals
D. None of the above.

_____ 17. Many electric circuits in DDC are Class 2, meaning they are _____.
A. encoded for security
B. high-voltage analog
C. power-limited
D. All of the above.

_____ 18. *True or False?* Direct digital control can be used to manipulate compressor operation.

_____ 19. A local controller is used to provide _____ control for a specific system or piece of equipment.
A. independent
B. interdependent
C. Both of the above.
D. None of the above.

_____ 20. A device that is regulated by optimized start/stop is controlled to _____.
A. indicate its status
B. minimize energy use
C. start and stop based on a timer
D. All of the above.

_____ 21. *True or False?* A packaged centralized computer control is a system composed of components from several manufacturers.

45.5 Building Control Protocols

_____ 22. Which of the following statements regarding pneumatic control systems is *false*?
A. They were originally designed to operate only with the manufacturer's equipment.
B. Proprietary designs required technicians to be trained specifically for each system.
C. They were inexpensive to install.
D. The systems were not interconnected and could not communicate with each other.

_____ 23. *True or False?* BACnet is a set of rules for hardware and software communication that allows controllers from one company to work with those from another company.

_____ 24. What device is used to connect networks within a system as well as to connect nonnative controllers and devices to the system?
A. Gateway
B. Router
C. Server
D. Workstation

12

_____ 25. Workstations that send instructions to and receive information from servers may be referred to as _____.

   A. clients
   B. controlled devices
   C. gateways
   D. routers

_____ 26. *True or False?* LonTalk is a communication protocol that permits integration of components from different manufacturers.

_____ 27. Modbus was originally developed for use with _____.

   A. existing pneumatic systems
   B. industrial hydraulic systems
   C. programmable logic controllers (PLCs)
   D. telecommunication equipment

45.6 Building Control System Diagnostics and Repair

_____ 28. The best way to start troubleshooting a building control problem is to begin with _____.

   A. a macro view of the system as a whole
   B. manually actuating each controlled device
   C. testing individual controlled devices (actuators)
   D. tracing the wiring between each controller and its devices

_____ 29. When troubleshooting a building Ethernet/IP problem, a technician should do the following, *except* _____.

   A. check Ethernet port LEDs for communication errors
   B. check network connectivity alarms
   C. run diagnostic software
   D. use a cable analyzer on each twisted pair of wires

_____ 30. When troubleshooting a controller and cable loss of signal problem, a technician should do the following, *except* _____.

   A. check for grounds between controller and controlled device
   B. check for shorts between controller and controlled device
   C. run diagnostic software
   D. use a cable analyzer on each twisted pair of wires

_____ 31. When troubleshooting to component level, a technician should do the following, *except* _____.

   A. analyze and test individual components suspected of problems
   B. measure current between sensors and controller
   C. run diagnostic software
   D. verify communication between controller and controlled device

_____ 32. Concerning electrical problems, if a contactor is buzzing but its power circuit is open, then _____.

   A. the armature is stuck
   B. a pressure switch is open
   C. the thermostat is open
   D. None of the above.

Name _____

_____ 33. If a contactor's power circuit is closed and a single-phase motor is humming but not running, _____.

    A. check its capacitor(s)

    B. the disconnect is open

    C. a fuse or circuit breaker is open

    D. the thermostat is open

_____ 34. If a contactor's power circuit is closed but a single-phase motor is not running or humming, _____.

    A. the contactor coil is shot

    B. measure motor for voltage

    C. a pressure switch is open

    D. the thermostat is open

Critical Thinking

35. List at least five factors that help determine energy use in a building.

_____

_____

_____

_____

_____

36. Explain why simple control systems are often more difficult to diagnose and repair than complex electronic control systems.

_____

_____

_____

_____

_____

12

Notes

CHAPTER 46

Energy Conservation

Name _____

Date _____ Class _____

> *Carefully study the chapter and then answer the following questions.*

46.1 Building Efficiency

_____ 1. A tool used to visualize and pinpoint places of heat loss in and around a building is best done using a(n) _____.

    A. anemometer
    B. duct blaster
    C. multimeter with amp clamp and thermocouple attachment
    D. thermal imaging camera

_____ 2. In the context of HVACR, *energy conservation* refers to a reduction in the _____.

    A. amount of energy needed for HVACR equipment operation
    B. run time allowed per day of HVACR equipment
    C. size of a building's conditioned space
    D. size of HVACR equipment for a building

_____ 3. One of the least expensive and easiest methods of conserving energy is to _____ a new or existing building.

    A. add a geothermal heat pump to
    B. earth shelter
    C. maximize the insulation in
    D. None of the above.

_____ 4. The following materials may be used as vapor barriers, *except* for _____.

    A. aluminum foil
    B. construction paper
    C. plastic film
    D. tarred paper

_____ 5. If insulation materials fill up with moisture, they will _____.

    A. form a corrosive slurry that will destroy nearby materials
    B. freeze solid in cold weather and chemically form a solid
    C. increase their insulating value
    D. lose much of their insulating value

12

_____ 6. The ability of insulation to resist heat transfer is known as its _____.

    A. I-value

    B. K-value

    C. R-value

    D. T-value

_____ 7. The US Department of Energy has developed guidelines for insulation based on _____.

    A. altitude

    B. climate zones

    C. occupant age

    D. occupant health condition

_____ 8. A concrete wall acts as a(n) _____, which retains heat for a long period of time.

    A. cold spot

    B. heat blanket

    C. thermal mass

    D. None of the above.

_____ 9. A building's envelope can be tightened by _____.

    A. installing high-performance windows and doors

    B. insulating and sealing gaps in the ductwork

    C. insulating and sealing the walls, floors, and ceiling

    D. All of the above.

_____ 10. A *degree-day* is the difference in the _____ temperature from the benchmark.

    A. actual average (mean)

    B. highest possible

    C. lowest possible

    D. None of the above.

_____ 11. The benchmark used in the degree-day method is _____.

    A. 32°F (0°C)

    B. 65°F (18°C)

    C. 90°F (32°C)

    D. 212°F (100°C)

_____ 12. Using the degree-day method, if a day's average temperature is 80°F, the day would have _____.

    A. 10 cooling degree-days in Fahrenheit units

    B. 10 heating degree-days in Fahrenheit units

    C. 15 cooling degree-days in Fahrenheit units

    D. 15 heating degree-days in Fahrenheit units

_____ 13. Using the degree-day method, if a day's average temperature is 50°F, the day would have _____.

    A. 15 cooling degree-days in Fahrenheit units

    B. 15 heating degree-days in Fahrenheit units

    C. 40 cooling degree-days in Fahrenheit units

    D. 40 heating degree-days in Fahrenheit units

Name _____

46.2 HVAC Equipment Efficiency

_____ 14. The Department of Energy's annual fuel utilization efficiency (AFUE) rating for oil-fired boilers must be _____.

    A. 55%

    B. 80%

    C. 95%

    D. 98%

_____ 15. A performance ratio that expresses a unit's cooling capacity in Btu/hr for each watt of power consumed is _____.

    A. annual fuel utilization efficiency (AFUE)

    B. coefficient of performance (COP)

    C. energy efficiency ratio (EER)

    D. seasonal energy efficiency ratio (SEER)

_____ 16. Air conditioners must have a SEER of _____ or higher to qualify for an Energy Star rating.

    A. 15

    B. 16

    C. 20

    D. 25

_____ 17. The heating or cooling effect per energy input under specific operating conditions is expressed by _____.

    A. annual fuel utilization efficiency (AFUE)

    B. coefficient of performance (COP)

    C. energy efficiency ratio (EER)

    D. seasonal energy efficiency ratio (SEER)

_____ 18. The efficiency of a heat pump's heating cycle must be measured using a COP or a _____.

    A. annual fuel utilization efficiency (AFUE)

    B. energy efficiency ratio (EER)

    C. heating seasonal performance factor (HSPF)

    D. seasonal energy efficiency ratio (SEER)

46.3 HVAC Alternatives for Energy Conservation

_____ 19. To improve the efficiency of a residential HVAC system's smaller single-phase motor fans, _____.

    A. equip them with variable frequency drives

    B. replace them with ECMs

    C. replace them with three-phase motors

    D. None of the above.

_____ 20. To improve the efficiency of a commercial or industrial HVAC system's fans, _____.

    A. equip them with variable frequency drives

    B. reduce their applied voltage in half

    C. replace them with ECMs

    D. None of the above.

12

_____ 21. Air-to-air heat exchangers that can be used to introduce fresh air to the conditioned space are _____.

    A. combustion chambers

    B. ERVs or HRVs

    C. subcoolers

    D. suction line-liquid line heat exchangers

_____ 22. A method of increasing a mechanical HVAC system's capacity and energy efficiency is to increase _____.

    A. ambient temperature

    B. head pressure

    C. subcooling

    D. superheat

_____ 23. _True or False?_ Evaporative cooling is best used in a humid climate.

_____ 24. A ponded roof is best used on _____.

    A. commercial one-story buildings

    B. four-story apartments

    C. skyscrapers

    D. None of the above.

_____ 25. The roof area of a ponded roof should be _____ the building's floor area.

    A. about half the size of

    B. as large as

    C. twice as large as

    D. None of the above.

46.4 The Role of the HVACR Technician

_____ 26. Maximum heat transfer is possible when a technician annually _____.

    A. cleans a system's evaporator and condenser

    B. explains EPA Clean Air Act requirements to customers

    C. measures suction and head pressures

    D. tightens all electrical terminal screws

_____ 27. Head pressure and compressor amp draw can be kept as low as possible by maintaining clean unobstructed airways _____.

    A. along the liquid line

    B. along the suction line

    C. around the step-down transformer

    D. at the condenser

_____ 28. Careful system maintenance includes _____ at every service call.

    A. adding acid to the refrigerant

    B. adding water to the refrigerant

    C. checking for refrigerant leaks

    D. venting the suction line

Name _____

Critical Thinking

29. Explain ways in which ductwork can contribute to energy waste.

_____

_____

_____

30. Describe five desirable traits of building insulation.

_____

_____

_____

_____

_____

31. Explain how the ventilation and insulation of vented attics contribute to the energy efficiency of a building.

_____

_____

_____

_____

32. Explain how blocking sunlight from an HVAC system's air-cooled condensing unit (while maintaining adequate space for airflow) would affect system efficiency, its electrical power consumption, its heat load, and subcooling. Include reasons for your explanation.

_____

_____

_____

_____

_____

33. Explain how a ponded roof works.

_____

_____

_____

34. Explain the ways that a technician can educate customers on reducing energy costs.

_____

_____

_____

_____

12

Notes

CHAPTER 47

Overview of Commercial Refrigeration Systems

Name _____

Date _____ Class _____

Carefully study the chapter and then answer the following questions.

47.1 Applications

_____ 1. Per FDA requirements, food storage equipment should maintain food temperature at or below _____ at all times.

A. 0°F (–18°C)
B. 15°F (–9°C)
C. 32°F (0°C)
D. 41°F (5°C)

_____ 2. An arrangement of compressors stored in the machinery room and piped in parallel with a common suction line, liquid line, and liquid receiver is a(n) _____.

A. blast chiller
B. ice bank
C. parallel compressor rack
D. split condenser

_____ 3. When intending to keep equipment as close together as possible, install a unit containing multiple compressors that circulate refrigerant through the evaporators in a nearby conditioned space. This type of unit is called a(n) _____.

A. distributed system
B. ice bank
C. locker plant
D. walk-in cabinet

_____ 4. A primary benefit of minimizing the distance between the separate parts of a commercial refrigeration system is a significant reduction of _____.

A. available floor space
B. control devices
C. refrigerant charge
D. refrigerated space

13

47.2 Commercial Refrigeration Systems

_____ 5. To prevent condensation from forming, commercial refrigerated cabinets commonly use ____ to warm certain outer edges.
  A. infrared lamps
  B. pilot flames
  C. resistance heating strips
  D. steam channels

_____ 6. Food freezers use a temperature alarm system to warn that temperature has reached ____.
  A. above an upper safe limit
  B. cut-in temperature
  C. freezing
  D. set point

_____ 7. Modern ice bank controls most commonly use a ____ to measure the thickness of the ice and control the level of liquids.
  A. cad cell and limit switch
  B. float valve and float switch
  C. membrane-split sensing bulb filled with water and a transmitting fluid
  D. set of stainless steel electrodes

_____ 8. Some walk-in cabinets may have a wall-mounted ____ that is separated from the main part of the cabinet interior by a vertical baffle.
  A. compressor
  B. condenser
  C. entry door
  D. evaporator

Match each of the components of the prefabricated walk-in drain installation with the proper term.

_____ 9. Drain cup with trap

_____ 10. Finish floor

_____ 11. Floor panel

_____ 12. Insulation

_____ 13. Nipple

_____ 14. Treated shims

_____ 15. Wall panel

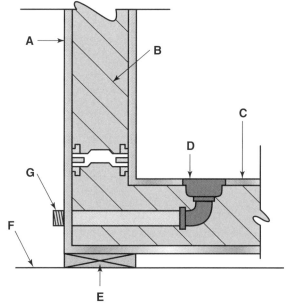

Goodheart-Willcox Publisher

Name _____

_____ 16. Most commercial display cases now use _____ evaporators for cooling.

    A. bare pipe

    B. forced-draft

    C. natural convection

    D. plate shelf

_____ 17. A misting system is most often used in display cases for _____ to prevent dehydration.

    A. dairy

    B. flower

    C. meat

    D. produce

_____ 18. In commercial refrigeration, an air curtain is used to _____.

    A. circulate conditioned air

    B. isolate a conditioned space

    C. move products along a line

    D. prevent products from moving

Match the components of the open frozen food case with the proper term.

_____ 19. Air curtain

_____ 20. Coil

_____ 21. Drain

_____ 22. Fan

_____ 23. Insulation

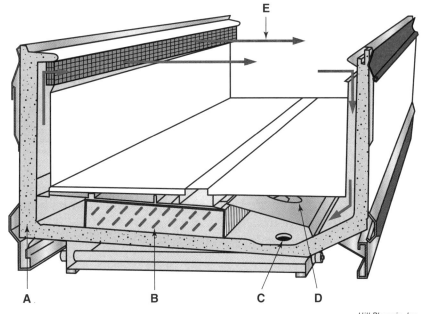

Hill Phoenix, Inc.

13

Match the components of the frozen food display case with the proper term.

_____ 24. Air curtain fan

_____ 25. Comfort zone jet

_____ 26. Condensate drain

_____ 27. Freeze jet

_____ 28. Guard jet

_____ 29. Insulation

_____ 30. Overhead light

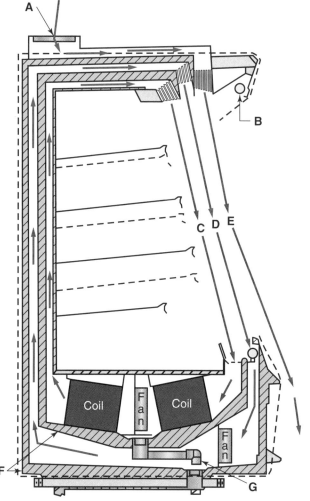

Kysor/Warren

_____ 31. According to the US Department of Agriculture, hot foods must be cooled down to _____ after two hours.

    A. 0°F (–18°C)

    B. 32°F (0°C)

    C. 41°F (5°C)

    D. 70°F (21°C)

_____ 32. According to the US Department of Agriculture, hot foods must be cooled down to _____ after four hours.

    A. 0°F (–18°C)

    B. 32°F (0°C)

    C. 41°F (5°C)

    D. 70°F (21°C)

Name _____

_____ 33. A system specifically designed to reduce the temperature of cooked food to refrigerated storage temperature is a _____.
  A. blast freezer
  B. hot and cold merchandiser
  C. quick chiller
  D. walk-in cabinet

_____ 34. A system specifically designed to reduce the temperature of cooked food to frozen storage temperature is a _____.
  A. blast freezer
  B. hot and cold merchandiser
  C. quick chiller
  D. walk-in cabinet

_____ 35. A beverage dispenser that transports syrup and water through stainless steel tubing surrounded by the ice that is dispensed for drinks is a(n) _____ chilled beverage dispenser.
  A. absorption
  B. mechanically
  C. passively
  D. None of the above.

_____ 36. A beverage dispenser that transports syrup and water through stainless steel tubing surrounded by a water bath cooled by a compression refrigeration system is a(n) _____ chilled beverage dispenser.
  A. absorption
  B. mechanically
  C. passively
  D. None of the above.

_____ 37. Since the cooling demand is irregular for water coolers, they must have a(n)_____.
  A. cryogenic phase of operation
  B. hold-over capacity
  C. shutoff valve in the drain line
  D. UV light to disinfect stagnant water

_____ 38. In a dispensing freezer, the temperature is kept within a narrow range of its set temperature and _____.
  A. cooled passively in tubing run through a reservoir of ice that is also dispensed for consumption
  B. the liquid mix is cooled or frozen under agitation
  C. primarily maintained by an ice bank produced during periods of low load
  D. the product is held in large plastic bags when ready for immediate consumption

_____ 39. To reduce the load on a bulk milk cooler, fresh milk is often passed through a(n) _____.
  A. cooling tower
  B. drizzler
  C. evaporative condenser
  D. plate heat exchanger

13

Match each of the components of the ice machine shown with the proper term.

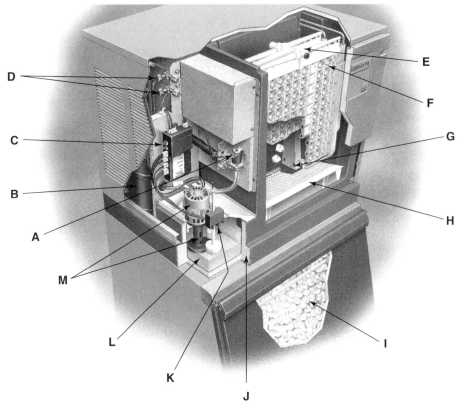

Scotsman Ice Systems

_____ 40. Cabinet insulation

_____ 41. Compressor

_____ 42. Control unit

_____ 43. Cube evaporator

_____ 44. Deflector tray

_____ 45. Float switch

_____ 46. Ice storage bin

_____ 47. Light curtain cube sensor

_____ 48. Service valves

_____ 49. Water distribution manifold

_____ 50. Water inlet solenoid valve

_____ 51. Water pump and motor

_____ 52. Water reservoir

_____ 53. An ice machine with a vertical cube evaporator senses when to switch from refrigeration cycle to harvest cycle by use of a(n) _____.

A. ice thickness sensor
B. light curtain
C. membrane-split sensing bulb
D. temperature sensor

_____ 54. An inverted ice cube evaporator _____.

A. continuously streams water across the molds to build up the ice
B. fills its cylindrical evaporator with water and freezes it
C. inverts its mold only after the ice has frozen
D. sprays cold water into its molds until ice is formed

Name _____

Match each of the components of the ice machine shown to the proper term.

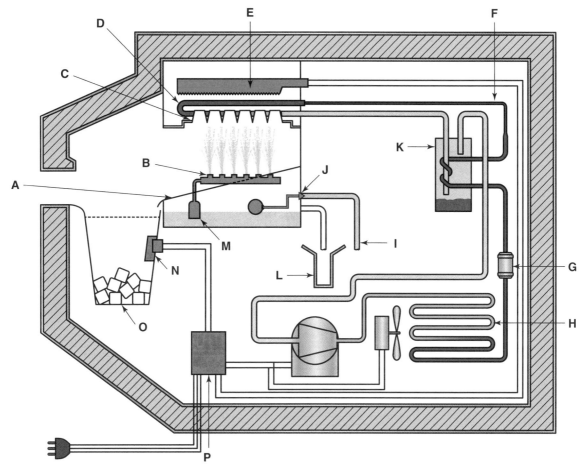

_____ 55. Accumulator

_____ 56. Capillary tube

_____ 57. Condenser

_____ 58. Controller

_____ 59. Deflector tray

_____ 60. Drain

_____ 61. Electric defroster

_____ 62. Evaporator

_____ 63. Filter-drier

_____ 64. Float valve/level control

_____ 65. Full bin shutoff control

_____ 66. Ice cube bin

_____ 67. Inverted ice cube mold

_____ 68. Spray nozzle manifold

_____ 69. Water pump

_____ 70. Water supply

_____ 71. A flaked ice evaporator _____.

    A. continuously streams water across the molds to build up the ice

    B. fills its cylindrical evaporator with water until ice builds up

    C. sprays cold water into its inverted molds until ice is formed

    D. uses a series of narrow honeycomb-shaped molds

13

Match each of the components of the ice machine shown to the proper term.

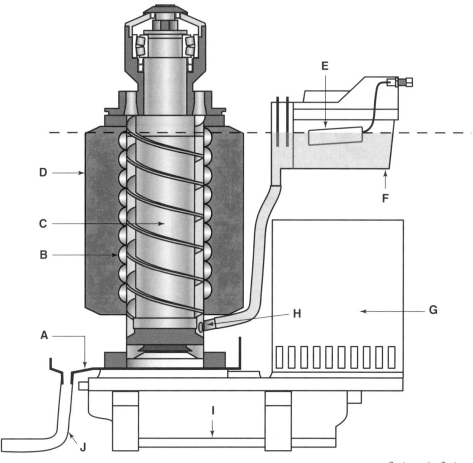

Scotsman Ice Systems

_____ 72. Auger

_____ 73. Auger motor

_____ 74. Drain

_____ 75. Drip pan

_____ 76. Evaporator insulation

_____ 77. Evaporator tubing

_____ 78. Gearbox

_____ 79. Level control

_____ 80. Water inlet

_____ 81. Water reservoir

47.3 Industrial Applications

_____ 82. An example of refrigeration in manufacturing and industrial processing includes the
following, *except* for the cooling of _____.

    A. compressed air for various processes

    B. electrodes on resistance welders using chilled water

    C. quenching liquids used to cool metals in heat-treating applications

    D. water into ice at a restaurant

Name _____

_____ 83. The primary reason why ordinary household refrigerators are not appropriate for storing flammable substances is that they _____.

    A. allow flammable vapors to escape too easily

    B. cannot reach the extremely low temperatures necessary

    C. have standard components that can cause a spark capable of igniting flammable vapors

    D. have a storage capacity that is too small

_____ 84. It is important to maintain high humidity in rooms where food is cured and stored, because this _____.

    A. allows meat to better maintain its taste and weight

    B. is easier on the refrigeration equipment

    C. prevents the growth of mold

    D. None of the above.

_____ 85. Cryogenic food freezing is most commonly done using liquid _____.

    A. argon and helium

    B. carbon dioxide and nitrogen

    C. carbon monoxide and oxygen

    D. neon and xenon

Critical Thinking

86. Briefly explain how an *ice bank* functions in a commercial refrigeration system.

_____

_____

_____

_____

_____

87. Refrigerated florist cabinets differ from refrigerated and frozen food display cabinets in several ways. List these differences and explain why they are necessary in a florist cabinet. Explain what these differences facilitate, prevent, or remedy.

_____

_____

_____

_____

_____

_____

88. Explain why all drain lines in refrigerated floral cabinets must be trapped.

_____

_____

_____

13

89. Briefly explain what happens when an open display case is exposed to significant drafts from unit heaters or fans. Explain why it is important to protect open display cases from drafts.

_____

_____

_____

90. The operation and efficiency of an ice machine with an air-cooled condenser is affected by the rising and falling of ambient temperature. Explain how a significant increase in ambient air temperature would affect ice production capacity and system efficiency. Also explain why this would occur.

_____

_____

_____

_____

_____

_____

91. Explain three methods used to remove ice from the freezing surfaces of different ice machines.

_____

_____

_____

_____

CHAPTER 48

Special Refrigeration Systems and Applications

Name _____

Date _____ Class _____

| Carefully study the chapter and then answer the following questions. |

48.1 Transportation Refrigeration

_____ 1. Medium-temperature and high-temperature refrigeration systems used in trucks and trailers typically use _____.

    A. R-22

    B. R-123

    C. R-134A

    D. R-404A

_____ 2. Low-temperature and medium-temperature refrigeration systems used in trucks and trailers typically use _____.

    A. R-22

    B. R-123

    C. R-134A

    D. R-404A

_____ 3. The system component that opens as needed to allow sufficient liquid refrigerant flow into the suction line to cool the compressor motor is the _____.

    A. hot-gas solenoid valve

    B. quench valve

    C. subcooler

    D. vibration absorbers

_____ 4. The system component that removes additional heat from refrigerant leaving the liquid receiver to help ensure that only liquid refrigerant enters the thermostatic expansion valve is the _____.

    A. hot-gas solenoid valve

    B. quench valve

    C. subcooler

    D. vibration absorbers

13

_____ 5. The system component that opens to allow refrigerant vapor discharged by the compressor to enter the evaporator is the _____.

    A. hot-gas solenoid valve
    B. quench valve
    C. subcooler
    D. vibration absorbers

_____ 6. The most common way of driving a compressor in a refrigerated trailer when the trailer is idle is power from _____.

    A. a diesel-powered generator
    B. the electrical grid
    C. a gear and linkage connected to the front wheels
    D. a generator connected to the radiator fan

_____ 7. The most common way of driving a compressor in a refrigerated trailer when the trailer is moving and in operation is power from _____.

    A. a diesel-powered generator
    B. the electrical grid
    C. a gear and linkage connected to the front wheels
    D. a generator connected to the radiator fan

Match each of the components of the refrigerated truck system with the proper term.

_____ 8. Clutch

_____ 9. Compressor

_____ 10. Condenser

_____ 11. Evaporator

_____ 12. Oil separator

_____ 13. Refrigerant lines

_____ 14. Remote controls

_____ 15. Vehicle battery

Carrier Transicold Division, Carrier Corps.

_____ 16. A thin, rectangular tank containing an evaporator surrounded by a solution that freezes at a desired temperature is a(n) _____.

    A. eutectic plate
    B. ice machine
    C. quench plate
    D. subcooler

Name _____

_____ 17. Since there may be inadequate airflow around the outside of refrigerated intermodal shipping containers on long trips, these systems often use ____.

A. cooling towers
B. keel coolers
C. water-cooled condensers
D. None of the above.

_____ 18. To cool a condenser using sea water, many boats use a(n) ____, which runs the condenser tubing outside the hull to take advantage of seawater's ability to absorb heat.

A. cooling towers
B. evaporative condenser
C. keel cooler
D. None of the above.

48.2 Alternative Refrigeration Methods

_____ 19. *True or False?* Low-pressure vapor nitrogen absorbs heat from the conditioned space as it condenses into low-pressure nitrogen liquid.

_____ 20. *True or False?* Expendable liquid nitrogen refrigeration systems may not require much power to operate, but they have intensive maintenance requirements.

_____ 21. Dry ice that is indoors at atmospheric pressure and at a temperature that is comfortable to humans will most likely ____.

A. combust
B. condense
C. melt
D. sublimate

_____ 22. *True or False?* Dry ice is sometimes used as an expendable secondary refrigerant.

_____ 23. Dry ice is composed of frozen solid ____.

A. carbon dioxide
B. carbon monoxide
C. nitrogen
D. oxygen

_____ 24. The most common dangers of touching dry ice with bare skin are ____.

A. blindness and fatigue
B. loss of consciousness and loss of memory
C. nausea and vomiting
D. severe burns and instant frostbite

_____ 25. When current is passed through the junction of two dissimilar metals, heat is absorbed in one part of the junction and moved to another part of the junction. This describes the ____ effect.

A. Bernoulli
B. Einstein
C. Peltier
D. Seebeck

13

_____ 26. *True or False?* Thermocouples are engineered to move heat when electric current is applied.

Match each of the components of the vortex tube assembly with the proper term.

_____ 27. Cold air out

_____ 28. Compressed air in

_____ 29. Deflector cone

_____ 30. Hot air out

_____ 31. Jet

_____ 32. Swirl chamber

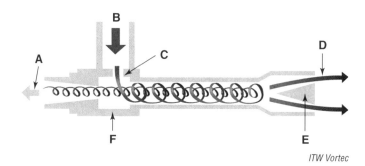

ITW Vortec

Match each of the components of the steam jet system with the proper term.

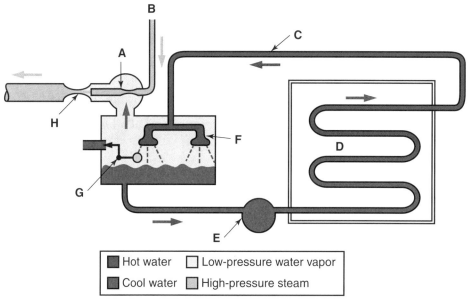

■ Hot water	□ Low-pressure water vapor
■ Cool water	▨ High-pressure steam

Goodheart-Willcox Publisher

_____ 33. Booster ejector

_____ 34. Circulating pump

_____ 35. Conditioned space

_____ 36. Spray nozzle

_____ 37. Steam line in

_____ 38. Steam nozzle

_____ 39. Water return

_____ 40. Water level control

Name _____

_____ 41. Refrigerant jet systems are mostly used on _____.
   A. installations with a large amount of waste heat
   B. intermodal systems for long distance
   C. small boats and yachts
   D. supermarket display cases

_____ 42. A *Stirling cycle* is a closed cycle that can convert _____ and vice versa.
   A. chemical energy into thermal energy
   B. electrical energy into chemical energy
   C. light energy into electrical energy
   D. thermal energy into mechanical energy

Critical Thinking

43. List five challenges faced in marine refrigeration design. Explain methods of meeting these challenges.

_____

_____

_____

_____

_____

_____

_____

44. In your own words, briefly explain how a steam jet system provides cooling.

_____

_____

_____

_____

_____

13

Notes

CHAPTER 49

Commercial Refrigeration—System Configurations

Name _____

Date _____ Class _____

Carefully study the chapter and then answer the following questions.

49.1 Commercial Systems Configuration Overview

_____ 1. Packaged systems _____.
   A. arrive charged, tested, and ready for installation
   B. arrive fully wired and piped
   C. include all of the major refrigeration components
   D. All of the above.

_____ 2. Refrigeration systems that are split are essentially _____ engineered.
   A. factory
   B. show room
   C. site
   D. None of the above.

49.2 Multiple-Evaporator Systems

_____ 3. Concerning a commercial refrigeration system with multiple evaporators, _____.
   A. each evaporator has its own thermostatic expansion valve
   B. the evaporators share a single thermostatic expansion valve
   C. a metering device is not used or needed
   D. Any of the above.

_____ 4. In a single commercial refrigeration system that controls two different evaporators at different temperatures, _____.
   A. each evaporator must have a check valve at its outlet
   B. an EPR is not used or needed at all
   C. an EPR is used on the evaporator at the lower temperature
   D. an EPR is used on the evaporator at the higher temperature

_____ 5. In a commercial refrigeration system with two evaporators, one of the evaporators must maintain a _____ in order to maintain a higher temperature than the other evaporator.
   A. higher airflow
   B. higher pressure
   C. higher refrigerant velocity
   D. lower airflow

13

Match each of the components of the commercial refrigeration system shown with the proper term.

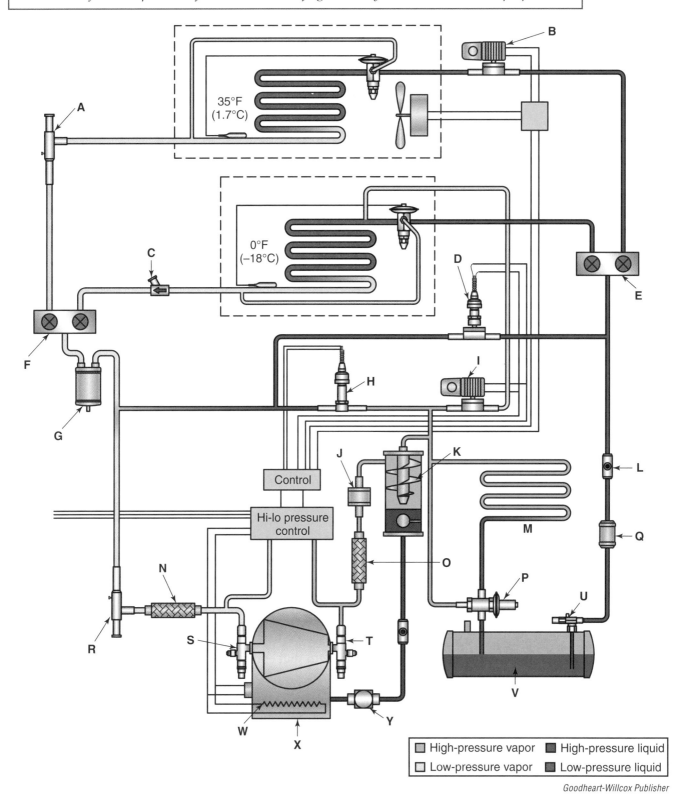

Name _____

_____ 6. Accumulator

_____ 7. Check valve

_____ 8. Compressor

_____ 9. Condenser

_____ 10. Crankcase heater

_____ 11. Crankcase pressure regulator

_____ 12. Discharge service valve

_____ 13. Discharge line vibration absorber

_____ 14. Evaporator pressure regulator

_____ 15. Filter-drier

_____ 16. Head pressure control valve

_____ 17. Hot-gas bypass valve

_____ 18. Hot-gas defrost valve

_____ 19. Liquid injection valve

_____ 20. Liquid line manifold

_____ 21. Liquid line solenoid valve

_____ 22. Liquid receiver

_____ 23. Liquid receiver service valve

_____ 24. Muffler

_____ 25. Oil return line shutoff valve

_____ 26. Oil separator

_____ 27. Sight glass

_____ 28. Suction line manifold

_____ 29. Suction line vibration absorber

_____ 30. Suction service valve

49.3 Modulating Refrigeration Cycle

_____ 31. *True or False?* The cooling capacity of a modulating refrigeration system is adjusted merely by turning the system *on* or *off*.

_____ 32. When there is nearly the maximum amount of heat to move in a multiple-compressor system, then _____.
  A. all the compressors cycle off
  B. all the compressors operate
  C. only one compressor runs
  D. None of the above.

_____ 33. When there is a very light and steady load and only a small amount of heat to move in a multiple-compressor system, then _____.
  A. all the compressors cycle off
  B. all the compressors operate
  C. only one compressor runs
  D. None of the above.

_____ 34. *True or False?* A refrigeration system's cooling capacity can be modulated by varying the speed of the compressor motor.

_____ 35. What type of circuit is often added to a system using hot-gas bypass as a form of capacity control?
  A. Dehumidifying circuit
  B. Desuperheating circuit
  C. Second stage circuit
  D. Subcooling circuit

13

_____ 36. A primary benefit of using hot-gas bypass as a method of capacity control is ____.

    A. less of a refrigerant charge

    B. lower electrical power usage

    C. precise humidity control

    D. None of the above.

_____ 37. When hot gas is bypassed into the suction line, instead of into the evaporator inlet, the system is likely to suffer from ____.

    A. burst sensing bulbs

    B. high head pressure

    C. high subcooling

    D. reduced proper distribution of oil

49.4 Multistage Systems

_____ 38. A multistage system has more than one stage of ____.

    A. compression

    B. condensation

    C. evaporation

    D. None of the above.

_____ 39. A compound refrigeration system has more than one compressor connected in ____.

    A. parallel

    B. reverse

    C. series

    D. None of the above.

_____ 40. The purpose of an intercooler in a compound refrigeration system is to ____ refrigerant.

    A. condense

    B. desuperheat

    C. evaporate

    D. subcool

Match each of the components of the compound system with the proper term.

_____ 41. Condenser cooling water in

_____ 42. Condenser cooling water out

_____ 43. Evaporator

_____ 44. High-stage compressor

_____ 45. Intercooler

_____ 46. Intercooler cooling water in

_____ 47. Intercooler cooling water out

_____ 48. Liquid receiver

_____ 49. Low-stage compressor

_____ 50. Oil return line

_____ 51. Oil separator

_____ 52. Temperature motor control

_____ 53. Temperature sensor

_____ 54. TXV

_____ 55. Water-cooled condenser

Name _____

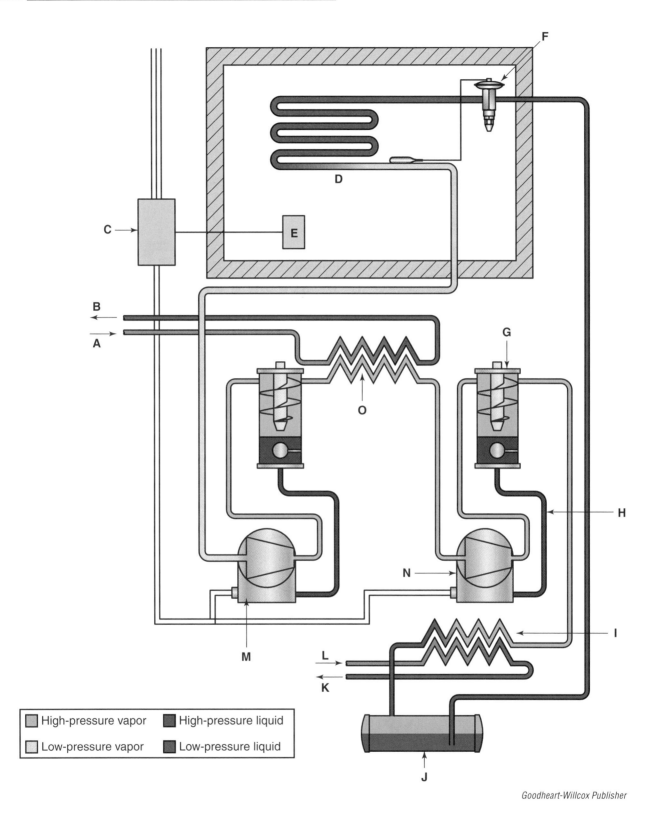

High-pressure vapor   High-pressure liquid

Low-pressure vapor   Low-pressure liquid

Goodheart-Willcox Publisher

13

_____ 56. *True or False?* Compound systems balance pressures during the Off cycle.

_____ 57. *True or False?* The motors of the compressors used in a compound system must be capable of starting under load.

_____ 58. A traditional mechanical refrigeration system maintains suction and head pressures during normal operation, and a compound refrigeration system maintains _____ in its refrigerant circuits during normal operation.
   A. only one pressure value
   B. two pressure values
   C. three pressure values
   D. four pressure values

Match each of the components of the cascade refrigeration system with the proper term.

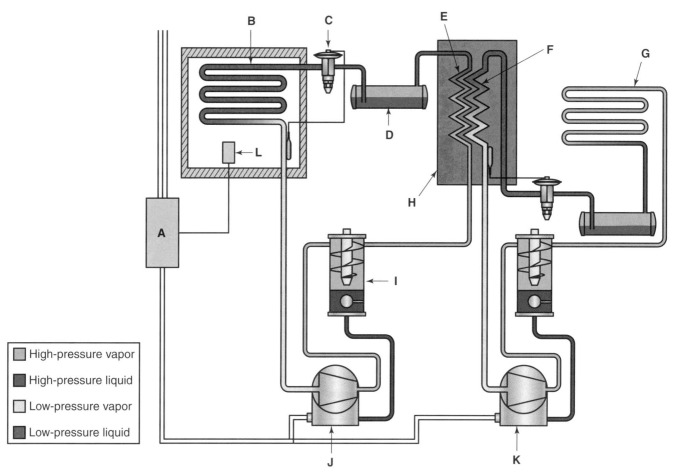

Goodheart-Willcox Publisher

_____ 59. Cascade heat exchanger

_____ 60. High-stage compressor

_____ 61. High-stage condenser

_____ 62. High-stage evaporator

_____ 63. Low-stage compressor

_____ 64. Low-stage condenser

_____ 65. Low-stage evaporator

_____ 66. Low-stage liquid receiver

_____ 67. Low-stage TXV

_____ 68. Motor control

_____ 69. Oil separator

_____ 70. Temperature sensor

_____ 71. *True or False?* The refrigerant in the high-stage subsystem is called the secondary refrigerant.

Name _____

49.5 Secondary Loop Refrigeration Systems

_____ 72. The fluid used in the secondary loop of a secondary loop refrigeration system differs from traditional refrigerant. However, the primary operational difference is that it _____.
   A. does not change phases
   B. is intended to conduct electricity
   C. is safe to consume
   D. moves through the circuit by magnetism

_____ 73. In a secondary loop refrigeration system, the primary loop is usually built with _____.
   A. black steel piping
   B. copper ACR tubing
   C. plastic piping
   D. stainless steel tubing

_____ 74. In a secondary loop refrigeration system, the secondary loop is built with _____.
   A. black steel piping
   B. copper ACR tubing
   C. plastic piping
   D. stainless steel tubing

_____ 75. The fluid flowing into the conditioned spaces of a secondary loop system is controlled by a _____.
   A. balancing valve
   B. float valve
   C. thermostatic expansion valve
   D. weight valve

_____ 76. The temperature of a conditioned space in a secondary loop system is primarily controlled by _____.
   A. air density
   B. fluid flow
   C. fluid pressure
   D. None of the above.

_____ 77. A cylinder with a pressure-responsive bladder that accounts for changes in secondary loop pressure describes a(n) _____.
   A. air separator
   B. expansion tank
   C. fill tank
   D. heat exchanger

_____ 78. Leaks in secondary loops are easy to detect because the fluid is _____.
   A. acidic
   B. dyed
   C. flammable
   D. pungent

13

Critical Thinking

79. Many commercial refrigeration systems operate multiple evaporators, and these evaporators can maintain different temperatures in their conditioned spaces. After their shared compressors cycle off, higher-pressure refrigerant could flow from one evaporator into another evaporator. Explain what would happen if higher-pressure refrigerant were to flow into a low-temperature evaporator. Explain how this might be prevented.

_____

_____

_____

_____

_____

80. Over the long term, explain what may result from a refrigeration system that short cycles frequently.

_____

_____

_____

_____

81. Briefly explain why the compressors in a multiple-compressor system are said to operate in parallel.

_____

_____

_____

_____

82. Is bypassing some of a compressor's discharge refrigerant the most efficient method of modulating system capacity? Explain your answer.

_____

_____

_____

_____

83. How does unloading one or more cylinders of a reciprocating compressor affect the refrigeration system's electrical power? Explain your answer. Also, explain how to measure and observe any changes to a system's electrical power. How would electrical measurements differ on a compressor when it is running all its cylinders and when it is unloading one or more cylinders?

_____

_____

_____

_____

_____

Name _____

84. Identify and briefly explain any problems that may occur when trying to reach high compression ratios using a single compressor.

_____

_____

_____

_____

85. The two subsystems of a cascade refrigeration system are said to be in series. Explain what this means.

_____

_____

_____

_____

_____

86. Explain why it is important to keep moisture out of cascade systems.

_____

_____

_____

_____

87. Explain some of the cost savings possible when using a secondary loop refrigeration system in a large supermarket installation instead of using a traditional refrigeration system.

_____

_____

_____

_____

13

Notes

CHAPTER 50

Understanding Heat Loads and System Thermodynamics

Name _____

Date _____ Class _____

| Carefully study the chapter and then answer the following questions. |

50.1 Heat Loads

_____ 1. *True or False?* If the inside of a refrigerated cabinet is cooler than the air outside of the cabinet, heat from the surrounding air will tend to transfer through the materials of the cabinet to the space inside.

_____ 2. Heat leakage into a refrigerated cabinet will increase as _____.

   A. temperature difference between inside and outside decreases
   B. temperature difference between inside and outside increases
   C. temperature equilibrium is nearly achieved
   D. None of the above.

_____ 3. Heat leakage into a refrigerated cabinet will increase as the cabinet wall _____.

   A. deteriorates
   B. thickness decreases
   C. thickness increases
   D. None of the above.

_____ 4. Heat leakage into a refrigerated cabinet increases the _____ its exterior area is.

   A. colder
   B. larger
   C. smaller
   D. None of the above.

_____ 5. The heat due to air changes, the temperature of articles put into the cabinet, additional heat brought into the cabinet due to operation, and heat generated inside the cabinet by fans, lights, and other electrical devices is accounted for by the _____.

   A. ambient heat load
   B. heat leakage load
   C. service heat load
   D. None of the above.

14

_____ 6. The total amount of heat that leaks through the walls, windows, ceiling, and floor of the cabinet per unit of time (usually 24 hours) describes the _____.
   A. ambient heat load
   B. heat leakage load
   C. service heat load
   D. None of the above.

_____ 7. *True or False?* The common unit used to measure surface area in heat flow calculations is the cubic inch.

_____ 8. A measure of how much heat can pass through one square foot of a material that is one inch thick in an hour when there is a temperature difference of 1°F (0.56°C) between one side of the material and the other describes the _____.
   A. C-value
   B. K-value
   C. R-value
   D. U-value

_____ 9. The measurement that takes into account the insulation value of boundary films is the _____.
   A. C-value
   B. K-value
   C. R-value
   D. U-value

_____ 10. The K-value and C-value for a material is equal when the material is exactly _____ thick.
   A. 1 mm
   B. 1 cm
   C. 1″
   D. 12″

_____ 11. Part of the service heat load includes removing _____ heat from fresh vegetables and meat.
   A. condensation
   B. perspiration
   C. respiration
   D. None of the above.

_____ 12. *True or False?* When the cold air inside a refrigerated space spills out through the bottom of the door opening, warmer room air moves into the refrigerated space.

_____ 13. *True or False?* A substance that is warmer than its surroundings will lose heat until the substance cools to the ambient temperature.

_____ 14. *True or False?* When the physical state of a material changes, its specific heat remains the same.

_____ 15. The following are variables that will affect the service heat load for a water cooler, *except* for the _____.
   A. rate at which water from the system will be consumed
   B. temperature of the people drinking the water
   C. temperature of the water entering the cooler
   D. temperature of the water leaving the cooler

Name _____

_____ 16. The amount of heat leakage in a water cooler is determined by the _____.

    A. exterior surface area of the materials used to construct the insulated water-storage parts

    B. materials used to construct the insulated water-storage parts

    C. thickness of the materials used to construct the insulated water-storage parts

    D. All of the above.

50.2 Thermodynamics of the Basic Refrigeration Cycle

_____ 17. The refrigerant that immediately vaporizes as it passes through the metering device is referred to as _____.

    A. bubble gas

    B. fast mist

    C. flash gas

    D. spray vapor

_____ 18. The degree of superheat is the difference between the refrigerant's evaporating temperature and the temperature of the vapor at the _____ inlet.

    A. compressor

    B. condenser

    C. evaporator

    D. liquid line

_____ 19. When a compressor compresses refrigerant vapor, it decreases the volume of the vapor, which in turn _____ of the vapor.

    A. decreases the temperature and pressure

    B. increases only the pressure

    C. increases only the temperature

    D. increases the temperature and pressure

Match each of the lines and points on the simplified pressure-enthalpy diagram with the proper term.

_____ 20. Constant heat line

_____ 21. Constant pressure line

_____ 22. Constant quality line

_____ 23. Constant temperature line

_____ 24. Saturated liquid line

_____ 25. Saturated vapor line

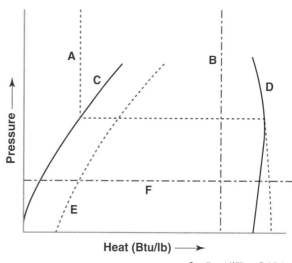

Goodheart-Willcox Publisher

14

Complete each sentence pertaining to the lines and points on the preceding simplified pressure-enthalpy diagram with the letter of the proper phrase.

_____ 26. Graph points that fall between lines C and D represent refrigerant that is _____.

_____ 27. Graph points to the right of line D indicate the refrigerant is _____.

_____ 28. Graph points to the left of line C indicate that the refrigerant is _____.

_____ 29. Along any vertical line, such as B, the refrigerant has _____.

_____ 30. Along any horizontal line, such as F, the refrigerant has _____.

_____ 31. Along stepped lines, like A, the refrigerant has _____.

A. all subcooled liquid

B. all superheated vapor

C. constant heat

D. a constant pressure

E. constant temperature

F. a mixture of saturated liquid and saturated vapor

_____ 32. A substance in vapor form that is in the presence of some of its own liquid is a _____ vapor.
  A. saturated
  B. subcooled
  C. superheated
  D. undercooled

_____ 33. *True or False?* The amount of latent heat contained in vaporizing and in condensing refrigerant changes at different pressures.

_____ 34. The term referring to a refrigerant's amounts of saturated liquid and saturated vapor is _____.
  A. refrigerant balance
  B. refrigerant quality
  C. refrigerant quantity
  D. refrigerant stratification

Name _____

> *Match the letter of each of the different pressure-enthalpy diagrams with its proper description.*

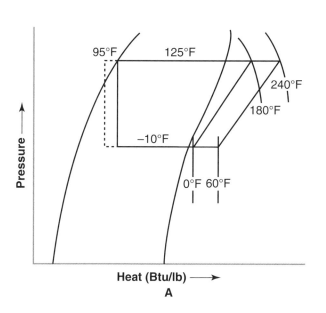

A

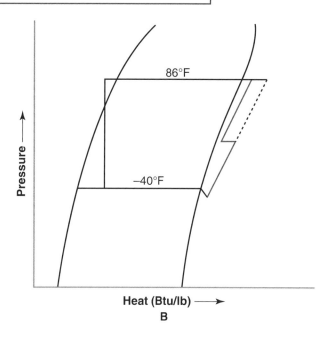

B

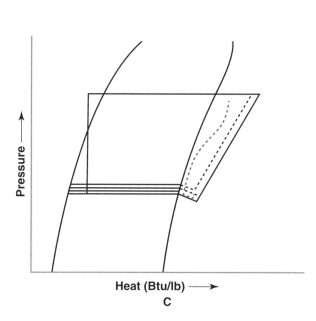

C

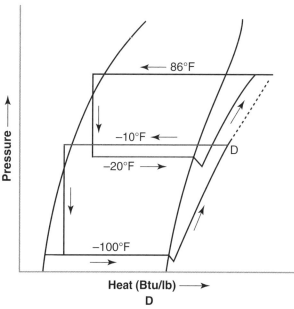

D

_____ 35. A refrigeration system using hot-gas bypass for capacity control.

_____ 36. A compound refrigeration system.

_____ 37. A cascade refrigeration system.

_____ 38. A refrigeration system with values of it with and without suction line insulation.

14

Critical Thinking

_____ 39. Felt has a thermal conductivity (K-value) of 0.25 Btu·in/hr·ft²·°F. Calculate the thermal conductance (C-value) of a 1/4″ thick layer of felt.

_____ 40. One layer of material in a wall has a thermal conductance (C-value) of 0.83 Btu/hr·ft²·°F. Calculate the thermal resistance (R-value) for this layer of material.

_____ 41. A composite wall has a total thermal resistance (R-value) of 5.4 hr·ft²·°F/Btu. Calculate the thermal transmittance (U-value) for the composite wall.

Name _____

42. Explain the four general steps a technician would use to calculate service heat loads using tables.

_____

_____

_____

_____

_____

_____

_____

_____

43. When a liquid product is going to be cooled below its freezing point, two sensible heat calculations must be performed. Explain why two calculations must be performed, instead of just one calculation.

_____

_____

_____

_____

44. Explain how the total heat released by a particular motor is calculated.

_____

_____

_____

_____ 45. Calculate the service heat load for a water cooler in an office building with 32 employees working a nine-hour shift. Note that the temperature of the supplied water is 65°F, and it must be chilled down to 50°F. Remember that one gallon of water equals 8.34 pounds, and the specific heat of water is 1 Btu/lb·°F. Assume that each person will consume 1/8 gallon per hour.

Notes

CHAPTER 51

Name _____

Date _____ Class _____

Commercial Refrigeration Component Selection

Carefully study the chapter and then answer the following questions.

51.1 Sizing Compressors, Condensers, and Evaporators

_____ 1. A commercial system's total heat load represents the amount of heat that the system must remove in a(n) _____-hour period.
   A. 1
   B. 10
   C. 12
   D. 24

_____ 2. When choosing a compressor for a commercial refrigeration system, relative humidity must be considered, because levels out of range can promote the growth of _____.
   A. insects and bugs
   B. mildew and mold
   C. rodents
   D. None of the above.

_____ 3. *True or False?* In some applications, a smaller compressor with a longer operating cycle can be as effective as a larger compressor with a shorter operating cycle.

_____ 4. The condensing temperature is the _____ temperature plus the temperature difference (TD).
   A. ambient design
   B. discharge
   C. liquid line
   D. suction line

_____ 5. The three factors that determine the capacity of a compressor are the following, *except* _____.
   A. condensing temperature
   B. refrigerant type
   C. suction line temperature
   D. type of lubricant

_____ 6. *True or False?* Total heat of rejection comprises the total heat load for the system and the energy added to the refrigerant by the compressor.

14

_____ 7. The three primary factors affecting a condenser's ability to transfer heat are the following, *except* _____.

    A. air/water velocity

    B. refrigerant quality (vapor and liquid mixture)

    C. surface area

    D. temperature difference

_____ 8. *True or False?* As the suction line temperature increases, the capacity of the evaporator increases to remove an increased amount of heat.

_____ 9. The initialism *FPI* stands for _____.

    A. (cubic) feet per inch

    B. fins per inch

    C. force per inch

    D. None of the above.

_____ 10. For air-cooling evaporators, the heat must pass through a(n) _____ on the evaporator tubing.

    A. air film

    B. layer of heavy paint

    C. layer of tubing insulation

    D. slime film

_____ 11. The primary factors that affect the heat transfer rate of a liquid-cooling evaporator are the following, *except* for _____.

    A. cooling water velocity

    B. evaporator construction

    C. tank insulation

    D. temperature difference

_____ 12. To account for a refrigerant circuit that could become restricted, a commercial refrigeration system's liquid receiver should be sized _____ larger than the total liquid volume in the system.

    A. 15%

    B. 48%

    C. 67%

    D. 100%

51.2 Calculating Theoretical Compressor Volume

_____ 13. The greater the _____, the lower the volumetric efficiency.

    A. head pressure

    B. low-side pressure

    C. valve opening

    D. All of the above.

_____ 14. Compressor efficiency will be reduced when extra pressure is generated in a reciprocating compressor's cylinder because its _____ valve is stuck.

    A. exhaust

    B. expansion

    C. intake

    D. relief

Name _____

_____ 15. The lower the _____, the lower the volumetric efficiency.

    A. head pressure

    B. heat of the compressor

    C. low-side pressure

    D. All of the above.

_____ 16. The greater the _____, the lower the volumetric efficiency.

    A. clearance space

    B. low-side pressure

    C. valve openings

    D. All of the above.

51.3 Designing Piping

_____ 17. The basic criteria that must be considered when sizing refrigerant lines are the following, *except* for _____.

    A. oil circulation

    B. pressure drop

    C. refrigerant line material

    D. refrigerant velocity

_____ 18. *True or False?* Both the length and diameter of refrigerant lines affect pressure drop.

_____ 19. The refrigerant line that should have a pressure drop of 0 psi and a velocity of 100 fpm or less is the _____ line.

    A. compressor discharge

    B. condenser condensate

    C. liquid

    D. suction

_____ 20. Excessive pressure drop can be caused by a refrigerant line that has too many _____.

    A. bends

    B. fittings

    C. valves

    D. All of the above.

_____ 21. Excessive pressure drop can cause _____ in a liquid line.

    A. excessive subcooling

    B. flash gas

    C. fractionation

    D. oil vaporization

_____ 22. The suction line should be sized so that the maximum pressure drop is equal to a saturation temperature drop of _____°F.

    A. 2

    B. 5

    C. 10

    D. 25

14

_____ 23. For horizontal suction lines, the refrigerant velocity is normally between ____.
   A. 500 fpm and 700 fpm
   B. 750 fpm and 900 fpm
   C. 1,000 fpm and 1,500 fpm
   D. 1,500 fpm and 2,000 fpm

_____ 24. For vertical risers in the suction line, the refrigerant velocity is normally between ____.
   A. 500 fpm and 700 fpm
   B. 750 fpm and 900 fpm
   C. 1,000 fpm and 1,500 fpm
   D. 1,500 fpm and 2,000 fpm

_____ 25. The refrigerant velocity in a horizontal run of the compressor discharge line is between ____.
   A. 150 fpm and 200 fpm
   B. 250 fpm and 400 fpm
   C. 500 fpm and 700 fpm
   D. 1,000 fpm and 1,500 fpm

_____ 26. Liquid refrigerant migration from the condenser to the compressor is possible during the Off cycle when the ____.
   A. compressor is warmer than the condenser
   B. condenser has a leak venting to atmosphere
   C. condenser is warmer than the compressor
   D. None of the above.

_____ 27. The compressor discharge line should be sloped toward the condenser a minimum of ____ per 10′ of run.
   A. 1/4″
   B. 4″
   C. 6″
   D. 12″

Name _____

> *Match each of the components of the riser configuration with the proper term.*

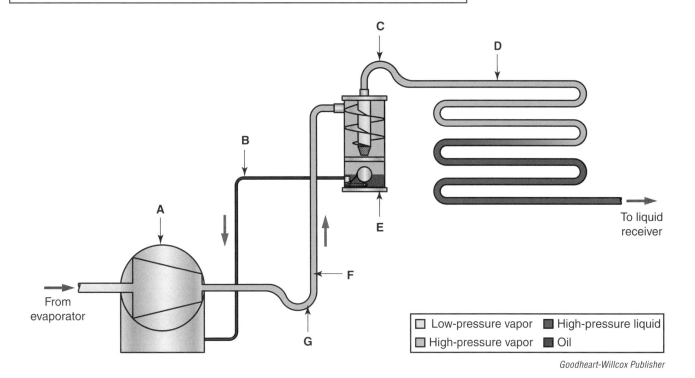

Goodheart-Willcox Publisher

_____ 28. Compressor

_____ 29. Condenser

_____ 30. Discharge line riser

_____ 31. Discharge line trap

_____ 32. Inverted trap

_____ 33. Oil return line

_____ 34. Oil separator

Critical Thinking

35. Identify four variables that affect the selection of the compressor, condenser, and evaporator.

_____

_____

_____

_____

14

36. Explain how the number of FPI on a condenser affects the condenser's heat transfer.

_____

_____

37. List at least four conditions that affect volumetric efficiency of a compressor.

_____

_____

_____

_____

38. Explain how compressor heat affects compressor efficiency.

_____

_____

_____

39. Explain what can result if not enough oil is returned to the compressor.

_____

_____

_____

_____

_____

40. Explain why vertical risers are a source of pressure drop in the liquid line.

_____

_____

_____

_____

41. Explain some of the installation methods a technician can use to counteract excessive pressure drop in the liquid line.

_____

_____

_____

_____

42. Explain how compressor capacity is reduced when there is excessive pressure drop in the suction line.

_____

_____

_____

_____

_____

Name _____

43. Explain why a technician should install an inverted trap at the top of a vertical riser in a suction line.

_____

_____

_____

_____

44. Explain why some modulating refrigeration systems use a double suction line in which one line is larger than the other line. How does this work with low-load and high-load conditions?

_____

_____

_____

_____

_____

_____

45. Explain some methods of preventing liquid refrigerant migration from the condenser to the compressor during the Off cycle.

_____

_____

_____

_____

_____

14

Notes

CHAPTER **52**

Installing Commercial Systems

Name _____

Date _____ Class _____

Carefully study the chapter and then answer the following questions.

52.1 Types of Commercial Installations

_____ 1. A slip hazard is most likely to result from a(n) _____ joint that has not been properly sealed.
  A. condensate line
  B. electrical conduit
  C. liquid line
  D. suction line

_____ 2. *True or False?* Due to the labor involved, it is more time-consuming to install a split system than a packaged system.

_____ 3. Two important principles of installation that help to minimize hazards are durability and _____.
  A. beauty
  B. inconspicuousness
  C. maximum speed
  D. neatness

_____ 4. When installing a new commercial system, a standing pressure test involves pressurizing the system with _____ and letting it stand for 24 hours.
  A. its intended refrigerant
  B. nitrogen
  C. oxygen
  D. water

52.2 Codes and Standards

_____ 5. Standards for wiring and electrical equipment are most often covered by _____.
  A. ASHRAE
  B. the IMC
  C. the IPC
  D. the NEC

_____ 6. Standards for water and other non-refrigerant piping are often covered by _____.
  A. ASHRAE
  B. the IMC
  C. the IPC
  D. the NEC

_____ 7. Most codes require the installation of commercial refrigeration equipment to be done by _____.
  A. anyone that doesn't get caught by the AHJ
  B. engineers
  C. general contractors with good liability insurance
  D. licensed refrigeration contractors

_____ 8. For each commercial refrigeration installation, a _____ must be obtained.
  A. bribe
  B. casual verbal agreement
  C. permit
  D. lookout

_____ 9. For commercial refrigeration systems, electrical work must conform to code and be done by a(n) _____.
  A. electrical engineer
  B. general contractor
  C. licensed electrician
  D. mechanical engineer

_____ 10. For commercial refrigeration systems, plumbing work must conform to code and be done by a(n) _____.
  A. electrical engineer
  B. general contractor
  C. licensed plumber
  D. mechanical engineer

_____ 11. Commercial refrigeration systems must be tested under pressure on the _____ and be free of leaks.
  A. high side only
  B. high side and the low side
  C. low side only
  D. None of the above.

52.3 Installing Condensing Units

_____ 12. For a large, split commercial refrigeration system, a condensing unit should be _____ the conditioned space.
  A. as close as possible to
  B. as far as possible from
  C. at least one floor above or below
  D. None of the above.

Name _____

_____ 13. If a condensing unit is placed indoors, the room must be ____.

    A. above grade
    B. below grade
    C. well lit 24/7
    D. well ventilated

_____ 14. A condensing unit should be installed indoors and ____.

    A. exposed to low or freezing temperatures
    B. exposed to the sun
    C. near steam pipes
    D. None of the above.

_____ 15. *True or False?* A high ambient temperature causes a condensing unit to operate less efficiently.

_____ 16. The safety device that is designed to actuate in the case of a fire is a ____.

    A. check valve
    B. fusible plug
    C. pressure control
    D. spring-loaded relief valve

_____ 17. The safety device that is used to stop the compressor when excessively high head pressure develops is a ____.

    A. check valve
    B. fusible plug
    C. pressure control
    D. spring-loaded relief valve

_____ 18. The safety device that is used to vent refrigerant only until high pressure is reduced is a ____.

    A. check valve
    B. fusible plug
    C. pressure control
    D. spring-loaded relief valve

52.4 Installing Expansion Valves

_____ 19. On a multiple-evaporator refrigeration system, a ____ should be included in each liquid line leading to each TXV.

    A. check valve
    B. pressure-regulating valve
    C. relief valve
    D. solenoid valve

_____ 20. The position of a TXV's sensing bulb on the suction line depends on the ____.

    A. evaporator temperature difference
    B. sensing bulb's refrigerant charge
    C. system's refrigerant charge
    D. tubing size

_____ 21. An effective way of ensuring that a TXV's sensing bulb is not affected by ambient air is to ____.

    A. coat the sensing bulb and its tubing section in thick grease

    B. install the entire sensing bulb within the suction line

    C. wrap the sensing bulb in insulation

    D. None of the above.

_____ 22. On a multiple-evaporator refrigeration system, different temperatures are maintained in the different evaporators primarily because of the proper operation of ____.

    A. CPRs

    B. EPRs

    C. solenoid valves

    D. TXVs

52.5 Installing Evaporators

_____ 23. To align and mark the proper placement of hangers for a wall-mounted or ceiling-mounted evaporator, ____.

    A. just eyeball it

    B. roughly estimate the spot

    C. use a cardboard template and a plumb line

    D. None of the above.

_____ 24. When an evaporator is mounted vertically, the evaporator outlet should be located ____ the inlet.

    A. above

    B. below

    C. level with

    D. None of the above.

_____ 25. For larger commercial evaporators, ____ is often used as the condensate tubing.

    A. cast iron

    B. hard PVC

    C. stainless steel

    D. All of the above.

_____ 26. In a multiple-evaporator system, a(n) ____ should be included between the suction line manifold and the outlet of the warmest evaporator.

    A. check valve

    B. crankcase pressure regulator

    C. evaporator pressure regulator

    D. solenoid valve

_____ 27. In a multiple-evaporator system, a(n) ____ should be included between the suction line manifold and the outlet of the coldest evaporator.

    A. check valve

    B. crankcase pressure regulator

    C. evaporator pressure regulator

    D. solenoid valve

Name _____

_____ 28. In a multiple-evaporator system, an accumulator is best installed _____.

A. between each evaporator and the suction line manifold
B. between each TXV and evaporator
C. in the main suction line near the compressor
D. in the main suction line near the evaporator

52.6 Installing Refrigerant Lines

_____ 29. The type of hard-drawn ACR tubing used in commercial refrigeration systems is usually Type _____.

A. K
B. L
C. M
D. O

_____ 30. Hard-drawn ACR tubing should be supported by clamps or brackets every _____ feet.

A. one to two
B. three to four
C. six to eight
D. fifteen to twenty

_____ 31. When setting aside unused tubing overnight, be sure to _____.

A. leave them in the middle of a walkway so you won't forget about them
B. pressurize them with a low-pressure mixture of nitrogen and refrigerant
C. remove any plugs or caps over the ends
D. secure plugs or caps over the ends

_____ 32. For tubing runs that cross a room, the tubing must be at least _____ above floor level.

A. 6"
B. 12"
C. 4'-6"
D. 7'-3"

_____ 33. *True or False?* Whenever possible, the inner parts of a valve should be removed while brazing the valve.

_____ 34. The most important criteria used to determine filter-drier size for a commercial refrigeration system are the following, *except* for _____.

A. compressor horsepower
B. compressor operating cycle (hours of operation per day)
C. tubing size
D. type and quantity of refrigerant

_____ 35. In commercial refrigeration systems, a sight glass should be installed between the _____.

A. compressor and the condenser
B. condenser and the liquid receiver
C. filter-drier and the metering device
D. liquid receiver and the liquid line

_____ 36. The best way to prevent the overheating of a moisture indicator when brazing the joints of a sight glass is to ____.

    A. direct flowing water through the sight glass

    B. remove the moisture indicator core

    C. submerge the sight glass in an ice bath

    D. use a fan to create a draft

52.7 Installing Electric Motors

_____ 37. _True or False?_ HVACR technicians are permitted to connect and service low-voltage wiring.

_____ 38. To hold mounting bolts in place until the motor is aligned and the washers and nuts are started on the bolts, some technicians use ____.

    A. caulking compound

    B. epoxy

    C. fishing line

    D. tack welds

_____ 39. The first thing that a technician should do before turning on any newly installed open-drive motor is ____.

    A. check for shorts to ground

    B. measure its current

    C. measure its voltage

    D. turn it by hand to see if it will rotate freely

_____ 40. When replacing a hermetic compressor, remember to ____ the joint after brazing tubing to the new compressor.

    A. apply flux to

    B. remove any remaining flux from

    C. use emery cloth on the outside of

    D. All of the above.

_____ 41. When working with wiring, avoid connecting copper and aluminum, because this will most likely result in ____.

    A. arcing and a carbonized connection

    B. burnt insulation that could lead to an electrical short

    C. galvanic corrosion

    D. spontaneous combustion

52.8 Testing Installations

_____ 42. When an inspector or code enforcement officer issues a ____ tag, it means that the job is complete and approved by the code authority.

    A. black

    B. green

    C. red

    D. yellow

Name _____

_____ 43. When an inspector or code enforcement officer issues a _____ tag, it means that work must stop until a plan to repair code violations is approved.

    A. black
    B. green
    C. red
    D. yellow

_____ 44. When an inspector or code enforcement officer issues a _____ tag, it means that work may proceed, but certain items must be brought up to code.

    A. black
    B. green
    C. red
    D. yellow

_____ 45. When leak testing a newly installed system using nitrogen, the pressure-relief valve should be adjusted to open _____ above the recommended testing pressure.

    A. 1 psi to 2 psi
    B. 20 psi to 25 psi
    C. 35 psi to 50 psi
    D. 100 psi to 150 psi

_____ 46. If a leak is determined to be in a brazed joint, _____.

    A. take it apart, reassemble it, and braze it
    B. use epoxy to patch the joint
    C. use mastic to patch the joint
    D. use an oxyacetylene torch to weld the joint

52.9 Charging Commercial Systems

_____ 47. *True or False?* During low-side charging, the high-side valve on the gauge manifold should be kept open.

_____ 48. Which position should the suction service valve be in during low-side charging?

    A. Back seated
    B. Front seated
    C. Mid-position
    D. None of the above.

_____ 49. *True or False?* Adding liquid refrigerant directly into the liquid receiver is best done with the compressor running.

_____ 50. According to how the liquid receiver service valve shown was installed, the valve should be in the _____ position to top off the refrigerant charge with the compressor running.

    A. back seated
    B. cracked open
    C. front seated
    D. None of the above.

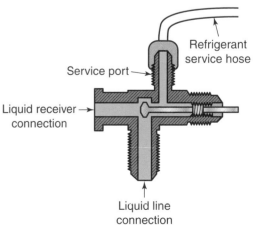

Refrigerant service hose

Service port

Liquid receiver connection

Liquid line connection

Goodheart-Willcox Publisher

52.10 Starting a Commercial Refrigeration System

_____ 51. When performing a hot pull down on a multiple-evaporator system, it is best to start with _____.
   A. all evaporator shutoff valves fully open
   B. all evaporator shutoff valves about mid-position
   C. one evaporator shutoff valve cracked open and all others closed
   D. the suction service valve closed for the first five minutes

_____ 52. When performing a hot pull down on a multiple-evaporator system by throttling the suction service valve (SSV), begin with the SSV so it is _____.
   A. back seated
   B. cracked open
   C. front seated
   D. in mid-position

_____ 53. When performing a hot pull down on a multiple-evaporator system, it is most important to monitor the compressor motor's _____ to ensure that it is not being overworked.
   A. continuity
   B. current draw
   C. resistance
   D. voltage

Critical Thinking

54. Explain why domestic refrigeration systems and other small-capacity systems are not usually covered or included in a municipality's codes.

_____

_____

_____

55. Explain why a flare nut used on a flare connection between an expansion valve and an evaporator in a commercial refrigeration system must be sealed after being secured in place. What could happen if the flare nut is not sealed?

_____

_____

_____

_____

56. Explain why sensing bulb mounting straps should be made of a material that conducts heat well.

_____

_____

_____

_____

Name _____

57. Explain why plastic straps should not be used to mount a sensing bulb to suction line tubing.

_____

_____

58. Explain why the suction line should be mounted with a slight slope toward the compressor.

_____

_____

_____

59. Explain why vibration absorbers are necessary in some commercial refrigeration system installations.

_____

_____

_____

60. Explain why it is recommended that low-side charging only be used to charge smaller commercial systems or to "top off" the existing charge in a system.

_____

_____

_____

61. Explain how to properly adjust refrigerant flow for a hot pull down when starting up a multiple-evaporator system using shutoff valves.

_____

_____

_____

_____

_____

_____

_____

_____

Notes

CHAPTER 53

Name _____

Date _____ Class _____

Troubleshooting Commercial Systems—System Diagnosis

> *Carefully study the chapter and then answer the following questions.*

53.1 Commercial Refrigeration Troubleshooting

_____ 1. In order for a system to operate correctly, the high side of the system must have enough ____.

    A. vapor pumped into the condenser at the correct pressure and temperature
    B. heat removed from the condenser
    C. vapor space (heat transfer surface area) in the condenser
    D. All of the above.

_____ 2. If there is a lack of refrigerant in a system, each pound of refrigerant will *not* completely ____.

    A. condense on the high side
    B. evaporate on the low side
    C. superheat before reaching the compressor
    D. None of the above.

_____ 3. *True or False?* When a system is undercharged, the total amount of latent heat absorbed into the evaporator refrigerant is increased over what an adequately charged system would absorb.

_____ 4. The most common noncondensable that gets trapped in a system is ____.

    A. air
    B. bits of broken up filter-drier element
    C. noble gases
    D. water

_____ 5. How does excessive head pressure electrically affect a refrigeration system?

    A. It causes arcing across contactor and relay contacts.
    B. It causes the compressor to draw less current.
    C. It causes the compressor to draw more current.
    D. It reduces the voltage across the compressor windings by half.

_____ 6. How does excessive head pressure affect condenser temperature?

    A. It does not affect condenser temperature.
    B. It lowers condenser temperature.
    C. It raises condenser temperature.
    D. None of the above.

_____ 7. How does excessive head pressure affect the temperature of the refrigerant fed to the refrigerant metering device?

A. It does not affect its temperature.
B. It lowers its temperature.
C. It raises its temperature.
D. None of the above.

_____ 8. How does excessive head pressure affect subcooling?

A. It decreases subcooling.
B. It does not affect subcooling.
C. It increases subcooling.
D. None of the above.

_____ 9. As a fluid's pressure decreases, it can _____.

A. absorb heat
B. expand to fill a larger volume and absorb heat
C. expand to fill a larger volume
D. All of the above.

_____ 10. Which of the following statements regarding moisture in a commercial system is *false*?

A. Any amount of moisture at or above the wet value ppm could be harmful.
B. The amount of moisture allowed will vary with refrigerant type and low-side temperature.
C. The color of the moisture indicator changes primarily with the variance in temperature and pressure.
D. Moisture indicators should be located downstream from the liquid line filter-drier.

_____ 11. If a filter-drier is clogged, it will feel _____.

A. cooler than normal
B. hotter than normal
C. no different than if it wasn't clogged
D. None of the above.

53.2 Checking Refrigerant Charge

_____ 12. The best and most precise way to ensure a proper refrigerant charge is to _____.

A. overcharge and drain excess through a petcock on the liquid receiver
B. overcharge and vent any excess from the low side
C. weigh in the exact prescribed amount
D. Any of the above.

_____ 13. If time and circumstances do not permit a technician to perform the superheat/subcooling methods, the refrigerant charge can be quickly checked by monitoring suction pressure, head pressure, and a liquid line's _____.

A. frosting
B. sight glass
C. solenoid valve
D. sweating

_____ 14. *True or False?* In all circumstances, bubbles in the liquid line indicate an insufficient charge in the system.

Name _____

53.3 Diagnosing Common Symptoms

_____ 15. Which of the following variables should be checked before a technician decides what the system's trouble is?

  A. Suction pressure and head pressure
  B. Superheat, subcooling, and the moisture indicator of the refrigerant
  C. The temperature of the evaporator, liquid line, and the suction lines
  D. All of the above.

_____ 16. If a unit using a capillary tube as its metering device produces insufficient refrigeration and is running continuously, most likely _____.

  A. low-side pressure will be higher than normal
  B. conditioned space temperature will be lower than normal
  C. an undercharge of refrigerant is the cause
  D. All of the above.

_____ 17. In an undercharged TXV system, suction pressure will be _____.

  A. higher than normal
  B. lower than normal
  C. normal
  D. None of the above.

_____ 18. Which of the following statements regarding an undercharged TXV system is *false*?

  A. The evaporator will be starved of liquid refrigerant and the refrigerant will quickly boil off.
  B. More heat than normal will be transferred to the high side.
  C. The refrigerant will have low to no subcooling.
  D. Some vapor may travel through the liquid line since there is less liquid refrigerant available.

_____ 19. A hot (not just warm) liquid line indicates that _____.

  A. the refrigerant has not condensed or subcooled
  B. there is a restriction in the discharge line
  C. there is a restriction in the liquid line
  D. there is a restriction in the suction line

_____ 20. A refrigeration system with a restriction in its capillary tube is likely to exhibit _____.

  A. high superheat
  B. higher than normal heat-absorbing capacity
  C. low superheat
  D. more refrigeration than normal

_____ 21. A refrigeration system with a restriction in its TXV is likely to exhibit _____.

  A. bubbles in the sight glass
  B. colder than normal conditioned space temperature
  C. high superheat
  D. low superheat

_____ 22. The simplest fix for a system with a restricted TXV that does not involve opening the refrigerant circuit involves _____.

    A. firmly striking the TXV valve body with a claw hammer
    B. melting any ice in the TXV
    C. rebuilding the compressor
    D. replacing the valve screen and filter-drier

_____ 23. *True or False?* A restriction in a TXV may reduce flow so much that temperature cannot drop enough to satisfy the thermostat during the On cycle.

_____ 24. If an expansion valve is stuck open, _____.

    A. the inlet screen may need to be replaced
    B. the valve may need to be flushed
    C. there may be dirt on the needle
    D. All of the above.

_____ 25. An inefficient compressor is most likely the cause of a system exhibiting _____.

    A. low suction pressure and high head pressure
    B. low suction pressure and low head pressure
    C. high suction pressure and low head pressure
    D. high suction pressure and high head pressure

_____ 26. When a compressor has become inefficient, _____.

    A. the evaporator will absorb more than enough heat to satisfy the thermostat
    B. some compressor vapor is pushed back into the low side through leaking inlet valves
    C. subcooling will very low or very high
    D. superheat will be low

_____ 27. An overcharged system using a capillary tube as its refrigerant metering device will exhibit low _____.

    A. head pressure
    B. subcooling
    C. suction pressure
    D. superheat

_____ 28. In an overcharged system using a TXV, the _____.

    A. bulk of the overcharge will remain on the low side
    B. overcharge raises the condenser head pressure but not its temperature
    C. subcooling will be lower than normal
    D. TXV will meter refrigerant into the evaporator based on superheat

_____ 29. When there is a restriction in the high side, the TXV will respond by opening _____.

    A. and closing fully in an erratic fashion to control the flow of refrigerant
    B. and closing to flush out its valve body
    C. wider to allow in more refrigerant to lower the subcooling
    D. wider to allow in more refrigerant to lower the superheat

_____ 30. If a restriction on the high side occurs at the inlet to the liquid receiver, then the system will exhibit the following, *except* _____.

    A. high head pressure
    B. high subcooling
    C. higher than normal condenser temperature
    D. low subcooling

Name _____

_____ 31. If a restriction on the high side occurs somewhere along the liquid line, then the system will exhibit the following, *except* _____.

A. higher than normal head pressure
B. lower than normal condenser temperature
C. lower than normal head pressure
D. normal subcooling

_____ 32. If temperature measurements at the condenser outlet tubing and the end of the liquid line are significantly different, the problem is most likely _____.

A. a restriction in the discharge line
B. a restriction in the liquid line
C. a restriction in the suction line
D. the TXV or refrigerant metering device

_____ 33. If a refrigeration system with a capillary tube as its refrigerant metering device has a restriction on its high side, the restriction will most likely be at the _____.

A. capillary tube inlet
B. compressor discharge valve
C. condenser inlet
D. None of the above.

_____ 34. A low-side restriction will _____.

A. cause refrigerant to accumulate in the evaporator
B. decrease evaporator temperature
C. lower conditioned space temperature
D. result in above normal superheat

_____ 35. When there is a low-side restriction in the suction line, the TXV will try to hold back liquid refrigerant in order to stabilize _____.

A. condenser temperature
B. evaporator temperature
C. subcooling
D. superheat

_____ 36. Faulty motor controls may be responsible for _____.

A. a compressor that short cycles
B. a compressor that will not start
C. a continuously running compressor
D. All of the above.

_____ 37. If a refrigeration system's condenser contains noncondensables, then head pressure will _____ after the system cycles off.

A. drop in a continuous, gradual manner
B. quickly and sharply drop
C. quickly and sharply rise
D. remain constant for several minutes

_____ 38. A refrigeration system with a very dirty evaporator will mostly exhibit _____.

A. high suction pressure and high head pressure
B. high suction pressure and low head pressure
C. low suction pressure and high head pressure
D. low suction pressure and low head pressure

_____ 39. If a refrigeration system will not cycle on, check compressor winding resistance using a(n) _____.

    A. clamp-on ammeter with power disconnected
    B. continuity tester with power connected
    C. ohmmeter with power disconnected
    D. voltmeter with power connected

_____ 40. If a compressor motor cycles off due to an internal overload, operation can begin again after _____.

    A. a cool-down period
    B. manually resetting the overload
    C. reprogramming the thermostat/system controls
    D. resetting the circuit breaker

_____ 41. The most relevant measurement to make after restarting a compressor that has cycled off due to an overload is _____.

    A. compressor current draw
    B. evaporator temperature differential
    C. subcooling
    D. superheat

_____ 42. A thermostat is most likely to cause a system to short cycle because its _____.

    A. differential is too large
    B. differential is too small
    C. sensing bulb is properly placed
    D. None of the above.

_____ 43. A low-pressure control is likely to short cycle on a compressor when a _____ is leaking.

    A. discharge service valve
    B. liquid receiver service valve
    C. suction service valve
    D. thermostatic expansion valve

_____ 44. The purpose of an anti–short cycle control is to prevent the compressor from _____ before a preset time has passed.

    A. cycling off
    B. cycling on
    C. satisfying the thermostat
    D. tripping the overload

_____ 45. When oil has been removed from a compressor, it should be tested for _____.

    A. acidity and contamination
    B. ignition temperature
    C. moisture content
    D. viscosity

_____ 46. The best option for pulleys that are causing noise is to be _____.

    A. annealed
    B. fitted with a custom counterbalance
    C. realigned
    D. All of the above.

Name _____

_____ 47. Once the rubber isolation absorbers of a condensing unit become brittle, the unit will _____.

A. arc and build up carbon on its electrical contacts
B. develop a short to ground
C. short circuit the compressor motor windings
D. vibrate disagreeably

53.4 Troubleshooting Ice Machines

_____ 48. An ice machine's water will develop a strong mineral concentration when the _____ drain becomes clogged.

A. bin
B. water-cooled condenser
C. water reservoir
D. None of the above.

_____ 49. To check the head pressure of an ice machine with a water-cooled condenser, the manufacturer usually provides a value that correlates with what temperature measurement?

A. Discharge line
B. Suction line
C. Water entering the condenser
D. Water leaving the condenser

_____ 50. As head pressure rises in an ice machine with a water-cooled condenser, the water valve should _____.

A. bypass most of its water directly to the drain
B. close completely
C. decrease water flow
D. increase water flow

_____ 51. If head pressure is within a manufacturer's recommended range but condenser water drain temperature is lower than normal, the system is showing signs of heat transfer that is _____.

A. increased between the evaporator and freezing water
B. increased between high-side refrigerant and condenser water
C. reduced between the evaporator and freezing water
D. reduced between high-side refrigerant and condenser water

_____ 52. *True or False?* When performing an ice capacity check on an ice machine with a water-cooled condenser, the ambient air temperature should be measured and recorded.

_____ 53. When a technician determines that an ice machine's condenser water circuit has a thin layer of mineral deposits, the technician should _____.

A. flush the water circuit with a scale remover
B. purge the water circuit with nitrogen
C. replace the condenser
D. replace the entire ice machine

_____ 54. *True or False?* When performing an ice capacity check on an ice machine, the technician must time the ice production cycle.

_____ 55. If the weight of the calculated daily ice production potential is within ____ of the manufacturer's recommendations, the machine is operating properly.

    A. 10%

    B. 20%

    C. 25%

    D. 37%

_____ 56. The thickness of ice cubes formed in a vertical evaporator ice machine is reduced by ____.

    A. adjusting the ice thickness sensor

    B. decreasing water flow

    C. increasing water flow

    D. installing a shielding cover that limits available room

_____ 57. *True or False?* Certain problems in flake ice machines can be easily heard due to the number of mechanical parts.

Critical Thinking

58. Explain at least two possible causes for pressure drop along a suction line.

_____

_____

_____

59. Explain how refrigerant is affected as it passes through a clogged filter-drier in the liquid line. How are pressure and temperature affected? Which gas law explains refrigerant behavior in this situation?

_____

_____

_____

_____

_____

60. A refrigeration system's sight glass can be used to check for bubbles. Explain several of the different causes of bubbles in a refrigeration system.

_____

_____

_____

_____

_____

Name _____

61. Explain how dirt or debris accumulation on an air-cooled condenser's fins and tubing affects efficiency. Explain how dirt and debris affect head pressure and condenser temperature.

_____

_____

_____

_____

62. Explain how head pressure and condenser temperature are affected when an air-cooled condenser's fan motor burns out. How will the system operate after this?

_____

_____

_____

_____

63. In a refrigeration system with a very dirty evaporator, explain why evaporator temperature is lower than normal but conditioned space temperature is higher than normal.

_____

_____

_____

_____

_____

64. Briefly explain how loose contaminants in the refrigerant circuit may cause a low-pressure control to short cycle the compressor back into operation.

_____

_____

_____

_____

65. Explain why it is best to avoid connecting pressure gauges to critically charged ice machines.

_____

_____

_____

_____

66. Explain the function of a water level/conductivity probe in an ice machine.

_____

_____

_____

_____

_____

_____

67. Explain why it is important to clean an ice machine's air-cooled condenser and its air filter regularly. What might happen if this maintenance practice is not followed?

_____

_____

_____

_____

68. Briefly explain what is involved in an ice machine's capacity check.

_____

_____

_____

_____

69. For an ice machine with an air-cooled condenser, explain how ambient air temperature affects a capacity check. What influence does ambient air temperature have on an ice machine's daily ice production?

_____

_____

_____

_____

_____

_____

70. For an ice machine, explain how supply water temperature affects a capacity check. What influence does supply water temperature have on an ice machine's production?

_____

_____

_____

_____

_____

Name _____

71. Explain what may occur when scale and deposits form on the spray nozzles of an inverted evaporator ice machine.

_____

_____

_____

72. What happens when the float switch in a flake ice machine is calibrated too high? Provide explanations for a float that is slightly too high and a float that is much too high.

_____

_____

_____

_____

_____

_____

_____

Notes

CHAPTER 54

Name _____

Date _____ Class _____

Troubleshooting Commercial Systems—Component Diagnosis

> *Carefully study the chapter and then answer the following questions.*

54.1 General Inspection Overview

_____ 1. To avoid overlooking details, a commercial refrigeration system should be inspected in a _____ manner.
   A. capricious
   B. careless
   C. random
   D. systematic

_____ 2. The following checks should be done on many open-drive compressors, *except* for _____.
   A. belt alignment
   B. belt condition and tightness
   C. coupling connection and condition
   D. water level

54.2 Checking Electrical Circuits

_____ 3. A major concern for electrical wiring for refrigeration systems is that the wires must be large enough in order to _____.
   A. carry the necessary amount of current
   B. handle the high voltage level
   C. provide sufficient resistance
   D. None of the above.

_____ 4. Before checking whether the proper voltage is available at a specific component, be sure to _____.
   A. lock out and tag out that circuit
   B. pull that circuit's fuse
   C. remove that component from the circuit
   D. set your meter to the proper setting

_____ 5. When using a clamp-on ammeter to check the current draw on a given electrical load, be sure to _____.

   A. clamp around only one of the load's wires
   B. electrically isolate the load from the circuit
   C. plug the meter leads into the correct jacks
   D. pull that circuit's fuse

_____ 6. When servicing a single compressor, do not start the unit without having a fully functioning _____ in the circuit.

   A. ignition transformer
   B. overload cut-out device
   C. rollout switch
   D. sequencer

52.3 Checking External Motors

_____ 7. A motor that is exhibiting excessive endplay can be remedied using _____.

   A. dielectric spray
   B. larger mounting bolts
   C. rubber mounting grommets
   D. washers

_____ 8. A noisy external motor should prompt a technician to check pulley bearing _____ to determine whether the bearings should be lubricated or replaced.

   A. conductivity
   B. current
   C. temperature
   D. voltage

_____ 9. A safe installation includes having a motor's _____ grounded.

   A. bearings
   B. frame
   C. shaft
   D. windings

_____ 10. An external motor's sheave should be checked using a(n) _____.

   A. anemometer
   B. draft gauge
   C. multimeter
   D. wear gauge

_____ 11. The most common trouble of the shaded-pole motors used to drive fans is worn _____.

   A. bearings
   B. fan blades
   C. relay contacts
   D. wiring terminals

Name _____

52.4 Checking Condensing Units

_____ 12. When an open-drive reciprocating compressor is producing a sharp clicking noise during operation, the problem is most likely _____.

    A. excess oil

    B. faulty inlet and discharge valves

    C. faulty service valves

    D. an overload that is stuck shut

_____ 13. Erratic refrigeration and constant oil slugging or pumping most likely indicates _____ in the system.

    A. excess oil

    B. insufficient oil

    C. the wrong type of oil was used

    D. None of the above.

_____ 14. An open-drive reciprocating compressor should be repaired or replaced if when it is isolated from the rest of the system, it cannot _____.

    A. hold and maintain pressures produced

    B. produce sufficient high pressure

    C. produce sufficient vacuum

    D. All of the above.

_____ 15. A gauge manifold is connected to an open-drive compressor that is isolated from the rest of the system. A technician turns the compressor's crankshaft a few times to produce vacuum in the crankcase and high pressure in the discharge. A crankshaft seal leak would then be indicated by _____.

    A. a gradual drop on the compound gauge

    B. a gradual rise on the compound gauge

    C. a gradual rise on the high-pressure gauge

    D. steady vacuum on the compound gauge

_____ 16. A worn piston or cylinder in an open-drive reciprocating compressor is indicated by a _____.

    A. dull clicking

    B. low humming

    C. sharp grinding

    D. shrill screeching

_____ 17. An open-drive reciprocating compressor's intake and exhaust valves can be checked for leaks by isolating the compressor from the system and connecting a gauge manifold. A leaking intake valve on the compressor is evident if the compound gauge measurement _____ after a vacuum is pulled.

    A. remains steady

    B. slowly drops lower

    C. slowly rises higher

    D. None of the above.

_____ 18. An open-drive reciprocating compressor's intake and exhaust valves can be checked for leaks by isolating the compressor from the system and connecting a gauge manifold. A leaking exhaust valve on the compressor is evident if the high-pressure gauge measurement _____ after being pressurized.

    A. remains steady
    B. slowly drops lower
    C. slowly rises higher
    D. None of the above.

_____ 19. Concerning any refrigeration system using a hermetic compressor, its _____ should be tested for acidity.

    A. evaporator condensate
    B. oil
    C. refrigerant
    D. All of the above.

_____ 20. The primary danger of acidity to a hermetic compressor is _____.

    A. acidic spontaneous combustion
    B. breakdown of winding insulation
    C. disintegration of metal parts
    D. leaks through the compressor shell

_____ 21. *True or False?* Most commercial compressors have an oil sump and oil drain to permit oil sampling and removal.

_____ 22. Commercial compressors often permit sampling and removal of oil through the use of an oil sump or _____.

    A. drain
    B. pressurized spigot
    C. scoop
    D. None of the above.

_____ 23. On a compressor motor that is in good working order, the normal winding resistance readings are usually _____.

    A. $0\ \Omega$
    B. in the single or low double digits
    C. in the high hundreds to low thousands
    D. infinity

_____ 24. When performing preliminary testing for grounded (shorted-to-ground) windings in a motor, a winding in good condition should show a measurement of _____.

    A. $0\ \Omega$
    B. in the single or low double digits
    C. in the high hundreds to low thousands
    D. infinity

_____ 25. A motor winding insulation test is best done using a(n) _____.

    A. capacitor tester
    B. megohmmeter
    C. ohmmeter
    D. thermal imaging camera

Name _____

_____ 26. The term *condenser capacity* generally refers to _____.

    A. the amount of heat a condenser can reject

    B. its maximum subcooling possible (at 68°F/20°C ambient temperature)

    C. its total surface area

    D. the volume of liquid refrigerant it can hold

_____ 27. How does surface area affect the amount of heat a condenser can reject?

    A. The more the surface area, the less heat it can reject.

    B. The more the surface area, the more heat it can reject.

    C. Surface area does not affect the amount of heat it can reject.

    D. None of the above.

_____ 28. How does ambient temperature affect the amount of heat an air-cooled condenser can reject?

    A. The higher the ambient temperature, the less heat it can reject.

    B. The higher the ambient temperature, the more heat it can reject.

    C. Ambient temperature does not affect the amount of heat it can reject.

    D. None of the above.

_____ 29. High head pressures in an air-cooled condenser may be caused by _____.

    A. low ambient temperature

    B. noncondensables in the system

    C. an undercharge of refrigerant

    D. None of the above.

_____ 30. *True or False?* Most commercial air-cooled condensers can be cleaned by a high-pressure jet of air, high-pressure water with detergent, mechanical scrubbing, or a vacuum cleaner.

_____ 31. Water flow through a water-cooled condenser should be adjusted so the temperature rise of the water does not exceed a _____ rise.

    A. 5°F

    B. 15°F

    C. 30°F

    D. 50°F

_____ 32. The materials sulfate, lime, and iron cling to the inner walls of a water-cooled condenser because of _____.

    A. an attraction of opposite charges

    B. a coating of sticky lubricant

    C. electromagnetism

    D. its textured surface

_____ 33. *True or False?* One indication of excess water flow is a continuous flow of water during the Off cycle in a water-cooled condenser.

_____ 34. A very noisy condition in which a single, distinct thump (rap) is heard in the pipes just as a valve closes describes the term _____.

    A. fluid strike

    B. liquid ram

    C. valve checking

    D. water hammer

54.5 Checking Liquid Lines

_____ 35. The primary purpose of a liquid line in a commercial refrigeration system is to deliver a steady stream of high-pressure ____.

    A. cold water
    B. superheated vapor refrigerant
    C. saturated refrigerant
    D. subcooled liquid refrigerant

_____ 36. Generally, the liquid line in a commercial refrigeration system should ____.

    A. be the same size as the liquid receiver service valve connection
    B. have reducer fittings
    C. have a temperature drop between the inlet and outlet of all its in-line components
    D. All of the above.

_____ 37. A quick way of locating possible unintentional pressure drops along a liquid line is by ____.

    A. installing a flowmeter at different points along the liquid line
    B. performing a temperature survey
    C. taking apart each joint for visual inspection
    D. All of the above.

54.6 Checking Thermostatic Expansion Valves (TXVs)

_____ 38. The placement of a TXV's bulb is based on the ____.

    A. size of the suction line tubing
    B. type of oil used in the system
    C. type of refrigerant in the sensing bulb
    D. type of refrigerant in the system

_____ 39. A starved evaporator would likely be caused by the following problems, _except_ for ____.

    A. a clogged inlet screen in a TXV
    B. frozen moisture in the TXV's valve body
    C. a loss of refrigerant from a TXV's sensing bulb
    D. a TXV's sensing bulb not being in contact with the suction line

_____ 40. Sweating or frosting on the suction line beyond a TXV's sensing bulb likely indicates ____.

    A. a high superheat setting
    B. insufficient refrigerant flow
    C. too much refrigerant flow
    D. None of the above.

54.7 Checking Electronic Expansion Valves (EEVs)

_____ 41. Most EEVs use a ____ to move and position the valve's plunger.

    A. manual adjustment screw
    B. pressure-responsive diaphragm
    C. sensing bulb and bellows
    D. stepper motor

Name _____

_____ 42. If an EEV's controller has a temperature sensor at the evaporator inlet and another temperature sensor at the evaporator outlet, it will be able to calculate and monitor _____.

    A. condenser saturation temperature
    B. discharge line superheat
    C. evaporator superheat
    D. subcooling

_____ 43. *True or False?* Controllers used with EEVs usually have a self-diagnostic function with codes specific to each manufacturer.

54.8 Checking Evaporator Pressure Regulators (EPRs)

_____ 44. Evaporator pressure regulators (EPRs) are known by these other names, *except* for _____.

    A. constant pressure valve
    B. open on rise of differential pressure (ORD) valve
    C. open on rise of inlet pressure (ORI) valve
    D. two-temperature valve

_____ 45. If a technician inspecting a multiple-evaporator refrigeration system noticed that an evaporator with an EPR was operating at a higher temperature than an evaporator without an EPR, it would appear that _____.

    A. everything seems to be operating properly
    B. there is a problem with the EPR
    C. there is a problem with the evaporator that does not have an EPR
    D. there is a problem with the evaporator that has an EPR

_____ 46. If a technician inspecting a multiple-evaporator refrigeration system compared two pressure measurements (one taken from an evaporator with an EPR and another taken from evaporator without an EPR), most likely the _____.

    A. evaporator with an EPR would have a higher pressure measurement
    B. evaporator without an EPR would have a higher pressure measurement
    C. pressure measurements would be equal
    D. None of the above.

_____ 47. If a technician took measurements that indicated that an evaporator was maintaining a cooler than normal temperature, the reason could be that the evaporator's _____.

    A. EPR inlet and orifice are both severely clogged
    B. EPR is leaking
    C. EPR is stuck shut
    D. None of the above.

_____ 48. If a technician took measurements that indicated that an evaporator was rather warmer than normal, the reason could be that the evaporator's _____.

    A. EPR is leaking
    B. EPR is stuck fully open
    C. EPR is stuck shut
    D. None of the above.

54.9 Checking Hot-Gas Valves

_____ 49. If there is a significant temperature difference between each tube connected to a hot-gas defrost valve during defrost operation, then _____.

    A. the compressor may be off

    B. the hot-gas valve may be stuck open

    C. the hot-gas valve may be stuck shut

    D. None of the above.

_____ 50. The least complicated type of hot-gas valve functions like a(n) _____.

    A. automatic expansion valve

    B. electronic expansion valve

    C. solenoid valve

    D. thermostatic expansion valve

_____ 51. For larger flow applications, a refrigeration system may use a(n) _____ valve.

    A. bellow-operated

    B. diaphragm-operated

    C. pilot-operated

    D. temperature-operated

54.10 Checking Solenoid Valves

_____ 52. The best position in which a solenoid valve should be mounted is _____.

    A. vertically with intended flow going downward

    B. vertically with intended flow going upward

    C. with the coil on bottom and the valve level

    D. with the coil on top and the valve level

_____ 53. A solenoid valve with dirty or loose electrical contacts is likely to _____.

    A. cause a compressor's motor to overload

    B. cause a short circuit and blow a fuse

    C. prevent the generation of enough electromagnetism to actuate the valve

    D. restrict fluid flow through the valve

_____ 54. If a solenoid valve is sticking, the technician should _____ the body of the valve to see whether this action allows hot gas to flow.

    A. disconnect

    B. evenly warm with a torch

    C. gently rap

    D. rotate

54.11 Checking Evaporators

_____ 55. Airflow through a direct-expansion evaporator can be checked with a(n) _____.

    A. anemometer

    B. gauge manifold

    C. multimeter

    D. superheat/subcooling calculator

Name _____

_____ 56. An air outlet or inlet that is _____ will result in inadequate cooling and abnormal temperatures and pressures.

A. too large
B. too small
C. unobstructed
D. None of the above.

_____ 57. As air passes through a direct-expansion evaporator, its temperature will usually drop about _____.

A. 15°F (8°C)
B. 30°F (19°C)
C. 45°F (25°C)
D. 50°F (28°C)

_____ 58. *True or False?* Simply measuring low-side pressure at the suction service valve reveals suction line restrictions or pressure drops.

_____ 59. The ideal superheat setting is when a TXV's sensing bulb temperature _____.

A. reacts directly to the air circulating through the evaporator
B. reacts to the system's subcooling
C. varies the least while the compressor is running
D. varies the most while the compressor is running

_____ 60. Pressure and temperature measurements on an evaporator often reveal symptoms that are the fault of the following, *except* for the _____.

A. EPR
B. evaporator
C. hot-gas valve
D. TXV

54.12 Checking Suction Lines

_____ 61. The primary purpose of a suction line in a commercial refrigeration system is to transport a steady stream of low-pressure _____.

A. cold water
B. superheated vapor refrigerant
C. saturated refrigerant
D. subcooled liquid refrigerant

_____ 62. An EPR is installed with the intention of maintaining its evaporator at _____ the suction line.

A. a higher pressure than
B. a lower pressure than
C. the same pressure as
D. None of the above.

_____ 63. A multiple-evaporator refrigeration system should have a pressure drop that is no more than _____.

A. 2 in. Hg
B. 2 psi
C. 14.7 psi
D. 30 psi

_____   64. When determining a suction line's pressure drop, it is necessary to take a pressure measurement ____.

   A. at the discharge service valve
   B. at the liquid receiver service valve
   C. downstream of an EPR
   D. upstream of an EPR

Critical Thinking

65. Explain why a motor should be running (not just starting or stopping) when measuring its voltage and current.

_____

_____

_____

_____

_____

_____

66. Explain why a technician should wait for a motor to cool down somewhat (not be hot) when checking it for open windings or an open circuit. Describe what might happen if a technician checked a hot motor for open windings or an open circuit.

_____

_____

_____

_____

_____

67. Explain how airflow affects an air-cooled condenser's heat rejection. What if there was no airflow? What if there was a lot of airflow? How would these two conditions affect the refrigerant in the condenser?

_____

_____

_____

_____

_____

Name _____

68. Explain how air trapped in the refrigerant circuit of an air-cooled condenser can affect the condenser's heat rejection.

_____

_____

_____

_____

_____

_____

_____

69. How does an *overcharge* of refrigerant affect head pressure? Explain the way in which an overcharge produces this effect.

_____

_____

_____

_____

_____

70. Explain how a technician can use large pieces of cardboard when checking an air-cooled condenser's operation. Also, explain why a technician would use cardboard in this way.

_____

_____

_____

_____

_____

71. Explain how pressure drop should be checked in a multiple-evaporator system.

_____

_____

_____

Notes

CHAPTER 55

Name _____

Date _____ Class _____

Servicing Commercial Systems

Carefully study the chapter and then answer the following questions.

55.1 System Service Fundamentals

_____ 1. Which of the following statements regarding pump-down or removal of refrigerant is *false*?

    A. The refrigerant can be pumped down into the liquid receiver instead of being removed from the system.

    B. Refrigerant may be removed from the system and stored in a recovery cylinder.

    C. Refrigerant must be removed from the part of the system that will be opened to the atmosphere.

    D. Refrigerant removal requires bringing the system into high pressure to evaporate the refrigerant from the section being dismantled.

_____ 2. The process of *atmospheric balancing* ends with the part of the system in question at _____.

    A. 0 psia

    B. 0 psig

    C. 14 kPa

    D. 14 psig

_____ 3. One of the safer and more reliable methods of achieving the proper pressure value for atmospheric balancing is done by _____.

    A. bypassing some refrigerant vapor through the gauge manifold

    B. cracking open a flare connection along the refrigerant circuit

    C. leaking some cylinder refrigerant to atmosphere

    D. loosening a hose connector at a service valve

Match each of the components of the gauge manifold with the correct term.

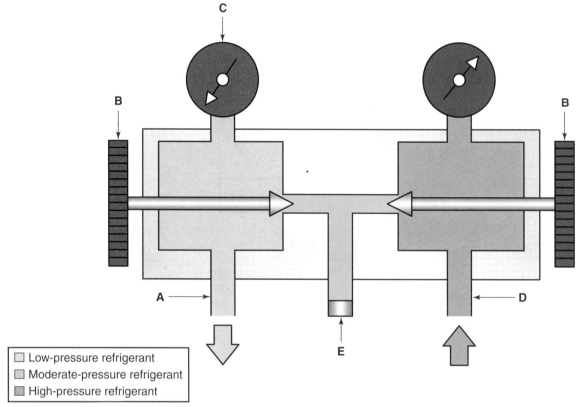

Low-pressure refrigerant
Moderate-pressure refrigerant
High-pressure refrigerant

Goodheart-Willcox Publisher

_____ 4. Center port (capped or closed by in-line hose valve)

_____ 5. Close valves when the compound gauge reads 0 psi

_____ 6. Connected to a high-side service valve that is only cracked open

_____ 7. Connected to a low-side service valve

_____ 8. Cracked open to throttle refrigerant

_____ 9. *True or False?* When refrigerant has been recovered into a cylinder, the vapor can be drawn back into the system from the cylinder, through the gauge manifold, and into the low side or high side of the system by having the system in vacuum.

_____ 10. When removing a compressor, evaporator, or TXV, the openings left in the tubing should immediately be _____.

    A. crimped and brazed closed
    B. filled deep with plumber's putty
    C. plugged with rubber stoppers
    D. threaded and screwed shut

Name _____

_____ 11. An open-drive compressor in a commercial refrigeration system has failed and needs to be replaced. The system does not have a liquid receiver, but there are service valves on the inlet and outlet of the compressor. You close the suction service valve, pump out the refrigerant within the compressor, and close the discharge service valve, so the compressor contains no refrigerant and is in vacuum. Next you should _____.

    A. disconnect the refrigerant lines from the compressor service valves
    B. disconnect the service valves from the compressor
    C. first A and then B
    D. first B and then A

Match the letter of each of the components shown with the correct term of this liquid recovery setup used on open-drive compressor systems.

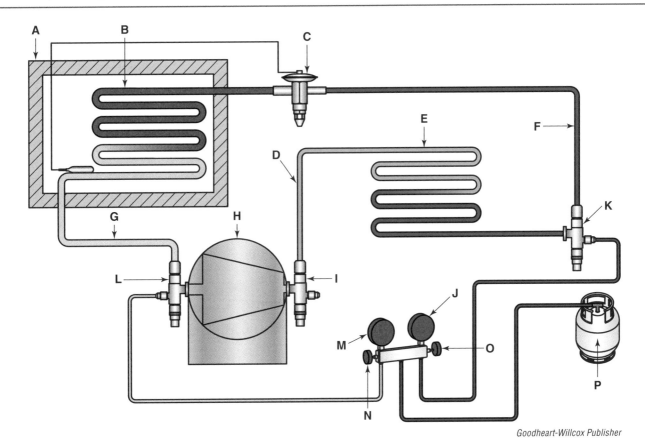

Goodheart-Willcox Publisher

_____ 12. Back seated	_____ 18. Front seated	_____ 23. Recovery cylinder
_____ 13. Compressor	_____ 19. Insulation	_____ 24. Suction line
_____ 14. Condenser	_____ 20. Liquid line	_____ 25. TXV
_____ 15. Cracked open off the back seat	_____ 21. Measuring head pressure	_____ 26. Valve open
_____ 16. Discharge line	_____ 22. Measuring suction pressure	_____ 27. Valve closed
_____ 17. Evaporator		

_____ 28. To prevent high head pressure and to minimize the time it takes when recovering refrigerant, a technician should _____.

    A. gently warm the recovery cylinder with heat lamps
    B. immerse the recovery cylinder in an ice water bath
    C. manually pump from the system and into the cylinder
    D. use only the vapor recovery method

_____ 29. Applying heat to a refrigeration system to reduce refrigerant recovery time can be done by _____.

    A. immersing the condenser and recovery cylinder in a solution of warm salt water
    B. quickly moving a torch with a very low flame over the condenser, liquid line, and evaporator
    C. using a heat lamp to warm the condenser, liquid line, and evaporator
    D. None of the above.

_____ 30. Large quantities of refrigerants recovered for disposal must be stored in _____ and moved to registered waste disposal sites.

    A. aluminum pouches
    B. containment bellows
    C. rugged cellophane bags
    D. steel drums

When reassembling a refrigeration system and returning it to service, actions must be done in the proper order. Match the letter of the step number with each step used.

_____ 31. Start and adjust the system for normal operation.   A. Step 1

_____ 32. Check for vacuum leaks.   B. Step 2

_____ 33. Clean and dry each part to be put into the system.   C. Step 3

_____ 34. Return all valves and controls to normal settings.   D. Step 4

_____ 35. Reintroduce refrigerant to the evacuated sections.   E. Step 5

_____ 36. After installing all necessary parts, evacuate that part of the system that had been opened.   F. Step 6

55.2 Servicing Motors and Compressors

_____ 37. *True or False?* When removing an open-drive compressor, start by draining the oil from the compressor and only stop once refrigerant starts escaping the drain hole.

_____ 38. When removing a belt-driven compressor for service, work within the following guidelines, *except* for _____.

    A. remove the flywheel before removing the compressor (if possible)
    B. rest the compressor weight on the flywheel
    C. set the down the compressor so that the flywheel is not touching anything
    D. All of the above.

Name _____

_____ 39. What is the primary reason that electric motors should be kept clean?

    A. Cleaning motors is easier than doing more difficult technical work.

    B. It follows the ABC rule: Always Be Cleaning.

    C. Keeping motors clean makes it easier to sell them.

    D. Most motors are air cooled and need their vents to be free and open for proper cooling.

_____ 40. Before removing an electric motor for service or replacement, _____.

    A. disconnect the power circuit, the power line, and remove the wires from the terminals

    B. label the terminals and wires for easier reassembly and loosen the motor's hold-down bolts

    C. remove any load from the shaft (flywheel, coupling, etc.)

    D. All of the above.

_____ 41. A burnout primarily involves the deterioration of _____.

    A. compressor gaskets

    B. discharge line tubing

    C. heat-responsive sensors and sensing bulbs

    D. motor winding insulation

_____ 42. When working with a burned-out compressor, _____.

    A. always wear goggles and rubber gloves to avoid skin contact

    B. do not allow the oil to contact your skin as it may be acidic

    C. wear a mask if necessary to avoid inhaling toxic fumes

    D. All of the above.

_____ 43. When servicing a system that had a hermetic compressor that burned out, be certain to install _____ after triple evacuation.

    A. a burnout compressor

    B. burnout filter-driers in the suction and liquid lines

    C. a burnout sight glass in the liquid line

    D. All of the above.

_____ 44. Refrigerant oil must meet the following requirements, *except* for _____.

    A. be compatible with the refrigerant in the system

    B. be designed to work properly in the low-side temperature range

    C. have the proper viscosity for the compressor

    D. have a wax content that is high enough for the system

_____ 45. When using a system's compressor to add oil to a refrigeration system, _____.

    A. attach the tubing equipped with a hand valve to the left opening of the gauge manifold

    B. draw a vacuum to draw oil into the system's liquid receiver

    C. draw a vacuum on the low side by turning the suction service valve all the way out

    D. evacuate the tubing connected to the gauge manifold, then immerse it in clean refrigerant oil

55.3 Servicing Condensers

_____ 46. Which of the following statements regarding air-cooled condenser removal is *false*?

A. The condenser should be thoroughly cleaned before it is removed.
B. If there is no shutoff valve between the condenser and the liquid receiver, the refrigerant should be pumped down and the condensate line crimped closed.
C. Liquid refrigerant must be recovered from the condenser or pumped down into the liquid receiver.
D. Residual oil in the condenser should be drained and measured upon removal.

_____ 47. Before an air-cooled condenser is removed, the condenser should be cleaned using the following methods, *except* for _____.

A. air, carbon dioxide, or nitrogen jets
B. a bleach- or acid-based solution
C. brushes
D. a vacuum cleaner

_____ 48. Which of the following statements regarding air-cooled condenser repair is *false*?

A. Before repairing a leak, the refrigerant tubes should be cleaned and flushed with nitrogen.
B. If a brazed joint is leaking, the outside should be cleaned. Apply flux to it, heat it, and then take apart the joint.
C. If any fins are bent, they should be straightened using a fin comb.
D. If a tube is cracked, the damaged part should be carefully pushed back into place and sealed with mastic.

_____ 49. A likely cause of high head pressure in a refrigeration system with a water-cooled condenser is _____.

A. high water flow rate
B. low water flow rate
C. a water valve stuck open
D. All of the above.

_____ 50. The most noticeable result of scale buildup in the water circuit of a water-cooled condenser is likely to be _____.

A. decreased heat transfer
B. increased heat transfer
C. reduced refrigerant flow
D. reduced water flow

_____ 51. Shut down and lock out all electrical service prior to opening the water circuit of a water-cooled condenser, otherwise _____.

A. the compressor motor will short out
B. the condenser fan could kick on
C. contact with water and the unit could lead to an electrical shock
D. the evaporator fan could kick on

_____ 52. When cleaning a water-cooled condenser, _____.

A. first shut down and lock out the electrical power
B. remove strainers and filters
C. flush and clean the valves and strainers and replace cartridge filters
D. All of the above.

Name _____

_____ 53. The usual way of ensuring that a water-cooled condenser's water circuit is free of any moisture is by ____.

    A. blowing out the water circuit with air, nitrogen, or carbon dioxide at 60 psig (415 kPa)

    B. flooding the water circuit with an inexpensive oil as a moisture displacer

    C. leaving the drain open for 5 to 10 minutes

    D. Any of the above.

_____ 54. *True or False?* When removing a water-cooled condenser that uses a circulating pump, the pump's drain plug should be left securely in place.

_____ 55. A technician that suspects that refrigerant is leaking into the water circuit of a water-cooled condenser should confirm such a suspicion using a(n) ____.

    A. anemometer

    B. bubble solution

    C. electronic leak detector

    D. flowmeter

_____ 56. Inadequate water flow through a water valve to a water-cooled condenser generally leads to ____.

    A. high head pressure

    B. low head pressure

    C. more efficient operation

    D. more favorable subcooling values

_____ 57. Excessive water flow through a water valve to a water-cooled condenser may be caused by the following, *except* for ____.

    A. high water pressure

    B. an improperly sized water valve

    C. low water pressure

    D. a poorly calibrated water valve

_____ 58. *True or False?* When a water valve needs to be replaced from a water-cooled condenser system, system operation and efficiency can be improved by simply installing a hand shutoff valve.

_____ 59. *True or False?* A thermostatic water valve that loses its element charge should be repaired, not replaced.

_____ 60. A pressure-operated water valve should be tested for leaks while being adjusted using a ____.

    A. dead weight tester

    B. flowmeter

    C. portable air cylinder

    D. water pump

_____ 61. *True or False?* No water should flow through a pressure-operated water valve until its control bellows pressure is reached.

_____ 62. *True or False?* The air passageways through a cooling tower should be washed to eliminate blockages and maximize airflow.

_____ 63. Tasks that should be performed on cooling towers on a monthly basis include the following, *except* for inspect _____.

  A. and clean air inlet screens
  B. pressures and pull a triple vacuum
  C. water level and adjust float, if necessary
  D. water for algae, leaves, or other dust particles

55.4 Servicing Liquid Lines

_____ 64. To change the direction of a run of hard-drawn ACR tubing, a technician should _____.

  A. anneal a small section of tubing and manually bend it
  B. install an angled brazed fitting
  C. use a hydraulic bender
  D. use a spring bender

_____ 65. Where should the sight glass and moisture indicator be installed along the liquid line?

  A. Between the compressor and the condenser
  B. Between the condenser and the liquid receiver
  C. Between the filter-drier and the refrigerant metering device
  D. Between the liquid receiver and the filter-drier

_____ 66. If a liquid line filter-drier is clogged or restricted in some way, _____ will appear in the sight glass.

  A. bubbles
  B. a color change
  C. a light
  D. nothing

_____ 67. If a liquid line is not just warm but noticeably hot, it probably contains _____.

  A. subcooled hot gas
  B. subcooled liquid
  C. superheated hot gas
  D. superheated liquid

_____ 68. On a commercial refrigeration system that has no sight glass, a restriction along the liquid line is located most quickly by performing a _____.

  A. deep vacuum
  B. dunk test
  C. leak test
  D. temperature survey

_____ 69. *True or False?* When restoring a system back to service after the liquid line has been evacuated, the technician should open the liquid receiver service valve as quickly as possible.

Name _____

55.5 Servicing Evaporators

_____ 70. Which of the following statements regarding evaporator removal is *false*?

A. All valves along the suction line must remain open so the pump down or recovery may continue unhindered.

B. Any hot-gas valves must remain open to prevent wasting time and effort pumping more than is necessary.

C. Any in-line components that may interfere with the pump down process should be identified and handled accordingly.

D. Pump down the system or recover the refrigerant charge.

_____ 71. Which of the following statements regarding system pump down is *false*?

A. If the compressor is being starved of oil, turn the system off and use a recovery machine.

B. The liquid receiver service valve must remain in mid-position while the compressor is running.

C. Unusual noises from the compressor indicate that the compressor is being starved of oil.

D. While the compressor is running, the technician should watch the compound gauge to see when atmospheric pressure is reached or a constant vacuum is produced.

_____ 72. If an ohmmeter measurement on an electrical defrost element isolated from its circuit reads OL or infinity (∞), then the technician knows that the element _____.

A. blew a fuse

B. is fine

C. has developed an open

D. has developed a short

_____ 73. Evaporator repairs are usually limited to the following tasks, *except* for _____.

A. flushing out the inside to remove scaling

B. locating and repairing leaks

C. repairing or replacing fins or motors

D. repairing or replacing fittings

55.6 Servicing Valves

_____ 74. When preparing to pump down a refrigeration system for the purpose of removing an expansion valve, it would make the most sense for a technician to close _____.

A. all suction line manifold valves

B. the discharge service valve

C. the nearest available valve upstream of the expansion valve

D. the suction service valve

_____ 75. After pumping down a refrigeration system for the purpose of removing an expansion valve, it would make the most sense for a technician to close _____.

A. all liquid line manifold valves

B. all suction line manifold valves

C. the discharge service valve

D. the nearest available valve upstream of the expansion valve

> *Match the letter of each of the components of this thermostatic expansion valve with the correct term.*

_____ 76. Adjustment screw

_____ 77. Bottom cap assembly

_____ 78. External equalizer connection

_____ 79. Inlet (liquid line connection)

_____ 80. Outlet (evaporator connection)

_____ 81. Push rods

_____ 82. Seal cap

_____ 83. Spring

_____ 84. Spring guide

_____ 85. Thermostatic element

Courtesy of Sporlan Division – Parker Hannifin Corporation

_____ 86. *True or False?* A clogged screen in a TXV's inlet could result in the starving of the evaporator.

_____ 87. *True or False?* To drive liquid refrigerant in the liquid line back to the nearest shutoff valve when preparing to remove an expansion valve, use an oxyacetylene torch with a generous blue flame to heat the liquid line.

_____ 88. *True or False?* If a new screen and gaskets are not available when a TXV is disassembled for service, it is best to just reinstall the TXV without an inlet screen or any gaskets and return the system to normal operation.

Name _____

_____ 89. If the evaporator temperature is too warm while the compressor is cycled on, the TXV should be adjusted to _____.

    A. allow more refrigerant into the evaporator
    B. prevent refrigerant flow
    C. reduce the refrigerant flow
    D. None of the above.

_____ 90. Evaporator pressure regulators (EPRs) _____.

    A. never have an access port anywhere on the valve body
    B. should be adjusted by turning the adjustment nut one full rotation at a time
    C. should be allowed 15 minutes of operation between each EPR adjustment
    D. None of the above.

55.7 Reconditioning Equipment after a Flood

_____ 91. When reconditioning equipment after a flood, be sure to take measurements along the entire electrical system using a(n) _____.

    A. capacitor tester
    B. clamp-on ammeter
    C. in-line ammeter
    D. ohmmeter

_____ 92. The outside of equipment that has been exposed to flooding should be cleaned with _____ before restarting.

    A. ammonia
    B. bacteria cleanser and detergent
    C. caustic acid
    D. fresh water

Critical Thinking

93. Explain what would happen if a technician did not perform atmospheric balancing before opening up part of a system's refrigerant circuit for service. What might be the long-term consequences of this action?

_____

_____

_____

_____

_____

94. When adding oil to a compressor using compressor suction, explain why it is important that some oil should be left in the glass container holding the filling tube at the end of the process.

_____

_____

_____

_____

95. Briefly explain how to add oil directly to a compressor crankcase without operating the compressor or using a pump.

_____

_____

_____

_____

_____

_____

96. Briefly explain why it is important to perform regular cleanings on air-cooled condensers.

_____

_____

_____

97. Briefly explain how to perform a "dunk test" on an air-cooled condenser. Also explain the purpose of a dunk test.

_____

_____

_____

_____

_____

98. Briefly explain why water-cooled condensers using well water require water treatment with water softeners and additional filters. Explain what would happen if no water treatment was used for the well water circulated through a water-cooled condenser.

_____

_____

_____

_____

_____

_____

Name _____

99. Constant water flow through a water valve into a water-cooled condenser can be wasteful and expensive. It can also obscure the presence of other problems. List two other negative effects of constant water flow. If the existence of these problems can be obscured by constant water flow, explain how.

_____

_____

_____

_____

_____

_____

_____

100. Briefly explain how a technician would balance the pressure in the evaporator to atmospheric pressure.

_____

_____

_____

_____

_____

_____

101. Explain why it is difficult to determine how best to move refrigerant and isolate different parts of the system in order to remove a hot-gas valve.

_____

_____

_____

_____

_____

_____

_____

_____

_____

Notes